THE SUSTAINABLE DEVELOPMENT PARADOX

THE SUSTAINABLE DEVELOPMENT PARADOX

Urban Political Economy
in the United States and Europe

Edited by
ROB KRUEGER
DAVID GIBBS

THE GUILFORD PRESS
New York   London

© 2007 The Guilford Press
A Division of Guilford Publications, Inc.
72 Spring Street, New York, NY 10012
www.guilford.com

Printed in the United States of America

This book is printed on acid-free paper.

Last digit is print number: 9 8 7 6 5 4 3 2 1

Library of Congress Cataloging-in-Publication Data

The sustainable development paradox : urban political economy in the United States
and Europe / edited by Rob Krueger, David Gibbs.
    p. cm.
Includes bibliographical references and index.
ISBN-13: 978-1-59385-498-0 (pbk. : alk. paper)
ISBN-10: 1-59385-498-6 (pbk. : alk. paper)
ISBN-13: 978-1-59385-499-7 (hardcover : alk. paper)
ISBN-10: 1-59385-499-4 (hardcover : alk. paper)
1. Sustainable development—United States.   2. Sustainable development—Europe.
3. Environmental policy—United States.   4. Environmental policy—Europe.
5. Regional planning—Environmental aspects—United States.   6. Regional planning—
Environmental aspects—Europe.   I. Krueger, Rob, 1968–   II. Gibbs, David, 1955–
  HC110.E5S875 2007
  338.973'07091732—dc22

                                                                    2007021363

Acknowledgments

The idea for this book originated before the 2004 annual meeting of the Association of American Geographers. We had put out a call for papers under the general title "Theorizing Sustainability." The response was overwhelming. We vetted over 40 abstracts and ended up with 20 papers and a panel discussion at the Philadelphia meeting. The presenters in the sessions generally focused on two clear themes: one that described individual cases of sustainability and sustainability as a form of resistance, and another that brought various political economic lenses to the practice of actually existing sustainability. It was clear from this experience—as well as subsequent ones at conferences in the United Kingdom, Europe, and the United States—not only that scholars of this orientation have a lot to contribute to these debates but also that it is absolutely necessary that they do so. We would like to thank the participants from these sessions and especially the contributors to this volume for their willingness to adopt these themes and for doing so adroitly. We would also like to acknowledge the valuable assistance of Ruth Anne McKeogh of Worcester Polytechnic Institute. She was an assiduous editorial assistant during the final phase of the project. Finally, we acknowledge the dedication and support of the staff at The Guilford Press, particularly Kristal Hawkins and Jeannie Tang. Without them this book would not have been completed.

Contents

Introduction: Problematizing the Politics of Sustainability   1
    Rob Krueger and David Gibbs

1 Impossible "Sustainability" and the Postpolitical Condition   13
    Erik Swyngedouw

2 Sustaining Modernity, Modernizing Nature: 41
    The Environmental Crisis and the Survival of Capitalism
    Roger Keil

3 Microgeographies and Microruptures: 66
    The Politics of Gender in the Theory and Practice
    of Sustainability
    Susan Buckingham

4 Containing the Contradictions of Rapid Development?:   95
    New Economy Spaces and Sustainable
    Urban Development
    David Gibbs and Rob Krueger

5 Greening the Entrepreneurial City?: Looking for   123
    Spaces of Sustainability Politics in the Competitive City
    Andrew E. G. Jonas and Aidan While

6 Integrating Sustainabilities in a Context of Economic, 160
Social, and Urban Change: The Case of Public Spaces
in the Metropolitan Region of Barcelona
Marc Parés and David Saurí

7 Political Modernization and the Weakening 192
of Sustainable Development in Britain
Anna Batchelor and Alan Patterson

8 Spatial Policy, Sustainability, and State Restructuring: 214
A Reassessment of Sustainable Community
Building in England
Mike Raco

9 The Spatial Politics of Conservation Planning 238
James P. Evans

10 The Imperial Valley of California: Sustainability, Water, 266
Agriculture, and Urban Growth
Stephanie Pincetl and Basil Katz

Index 299

About the Editors 307

Contributors 309

Introduction

Problematizing the Politics of Sustainability

ROB KRUEGER
DAVID GIBBS

SUSTAINABILITY IN CONTEXT

The discourse of sustainability is being more widely deployed as an urban and regional development strategy than ever before. ICLEI—Local Governments for Sustainability, formerly the International Council for Local Environmental Initiatives, reports, for example, that over 6,000 local governments from 113 countries have adopted sustainability initiatives since the early 1990s (International Council for Local Environmental Initiatives, 2002). Similarly, Jacobs (1999) reports that central governments are also promoting local and regional sustainability programs, especially in terms of the environment. Perhaps the most bold declaration regarding the "state of sustainability" comes from Campbell (1996: 312), who declared that "in the battle of big public ideas, sustainability has won: the task of the coming years is simply to work out the details, and to narrow the gap between its theory and practice." Indeed, it's really quite difficult to find anyone who isn't in favor of sustainability. As Erik Swyngedouw so poignantly observes in Chapter 1 of this volume:

"Greenpeace is in favor, George Bush Jr. and Sr. are, the World Bank and its chairman (a prime war monger in Iraq) are, the Pope is, my son Arno is, the rubber tappers in the Brazilian Amazon are, Bill Gates is, the labor unions are." And who wouldn't be in favor of sustainability? Social equity, environmental stewardship, and economic prosperity are all noble policy goals—regardless of one's political orientation. We are reminded of Richard Nixon's famous remark more than a quarter-century ago: "We are all Keynesians now." Has the idea of sustainability become similarly embedded in our culture now such that we might aptly call ourselves "sustainabilists," or ecological modernists? While such claims may make excellent political rhetoric, a more considered answer to this question requires rigorous engagement with the concepts and issues of social change.

Engaging the politics of sustainability represents a gap in the current sustainability literature. On one side, the project of sustainability in the academy is represented by the development of grand models, even down to describing particular practices. While these approaches represent the dichotomy between theory and practice, what they have in common is a belief in the power of the current system and faith in its intrinsic ability to regulate itself to produce progressive outcomes. These accounts are thus tinged with the neoclassical logic of "getting it right" and often conclude (or fizzle out) in a vague appeal to a mix of moral imperative, social conscience, and the threat of looming ecological disaster. The grander approaches, exemplified by the work of Costanza et al. (1997), Hawken, Lovins, and Lovins (2000), and Brown (2001), focus on reformulating economic systems, especially in terms of redefining how economic value is calculated. While these analyses are rigorous, especially those of ecological economists like Costanza, and may even provide short-term fixes to environmental problems, they return to the same propositions as their neoclassical predecessors: that the market, properly defined, incentivized, and reflecting real costs of development, is the most desirable institution for delivering human prosperity and ecological integrity. Here, market-produced values are surrogates for the value of "environment." Wastewater treatment plants, for example, can account for the ecological services of a wetland by calculating the cost of construction, maintenance, and human resource requirements. Thus, the economic value of a wetland is established by its ability to cleanse impurities from water. It's about getting the price right, full stop. Similarly, Daly's (1974) notion of steady-state economics suggests that economic systems should only grow at a rate such that natural resources can replenish themselves. Once growth surpasses replacement, disincentives should trigger slower growth. While these developments represent a breakthrough in valuing the environment in ways that traditional economists have not, the analysis falls short of engaging such key questions as: What "environment" is being saved? For whom? To what end? How do we measure a contested concept like "nature's" services? How do institutions work under conditions of economic change?

The other primary area where sustainability flourishes as an academic project is in the case study arena. Here, analysts have developed case studies seeking to show sustainability in action in various places and under a variety of policies. In theoretical terms this approach may be informed by the "new localism" literature, a viewpoint that asserts the efficacy of localities in promoting sustainability (see Selman, 1996; Marvin & Guy, 1997). Local Agenda 21 refers to this concept of the power and appropriateness of the local entity as "subsidiarity." This is the idea that local authorities are closest to the people and therefore most responsive to community calls for sustainability. Thus, under this view, case studies are useful because they share unproblematic information among various localities, a sort of menu of policy measures and processes from which local authorities can choose in order to implement their own forms of sustainable development. Thus, case studies offer a "pick and mix" set of policies—bicycle lanes here, high-density zoning there, urban green space preservation inside the city boundary, and improved public transport systems, for example—that can be adapted to local circumstances. Volumes have been written presenting practical examples of what such sustainable lifestyles might look like (see e.g., Beatley & Manning, 1997; Carley & Spapens, 1998; Pierce & Dale, 1999; Beatley, 2000; Calthorpe & Fulton, 2001; Portney, 2003). Cities including Portland, Oregon, Santa Monica, California, Freiburg, Germany, and Malmo, Sweden, as well as whole nations such as the Netherlands and Sweden—Scandinavia, for that matter—have been identified and praised for their efforts to bring models of sustainability for the rest of the world to follow.

Another source of inspiration for local sustainability and the reassertion of local primacy is the ecological footprint metaphor, developed by Wackernagel and Rees (1996). Ecological footprints seek to measure the impact that city living has on the earth. People, for example, require resources beyond that provided by their individual land holdings. The purpose of ecological footprint analysis, therefore, is to identify the number of acres/hectares humans living in particular cities require to exist. Often, however, the rhetoric of sustainability, ecological footprints, and carbon reduction strategies can get ahead of what is possible in these times. In the state of Massachusetts, for example, residents can now purchase energy generated from renewable sources (the background to which is the neoliberal deregulation of the utility industry in the United States), but this has impacts that are wider than the state level. While there are some small producers in the region, much of this renewable energy comes from Hydro-Quebec, which has social consequences for the Cree Nation in Canada that once obtained its traditional livelihood from lands that are now under water as part of the Hydro-Quebec system of electricity-producing dams. Sometimes in our efforts to be "green" we can trample on the basic human rights of others. Thus, when we think about sustainability, we have to think beyond our political borders, which extend from the body outward, to consider the implications of these decisions.

Recently scholars of sustainability have begun to examine sustainability as a more complex social process. In his book *Taking Sustainable Cities Seriously*, Portney (2003) suggests, at least in the U.S. context, that sustainability is too early in development and should not be assessed for outcomes yet, but rather whether it is moving along in the process. He focuses instead on the activities—policies, programs, organizations, and practices—that are utilized in U.S. cities (Portney, 2003: 2). Portney draws upon regime theory and urban growth coalitions, among other things, to problematize the "American process" of sustainability. Indeed, other work has also drawn upon Molotch's growth coalition work (Molotch, 1976) to describe and problematize urban sustainability, such as "smart growth," in the American context (Gearin, 2004). But these efforts remain highly disciplinary and focused on only one dimension of sustainability—the environment. Sustainability, however, is supposed to be a three-legged "stool"—not a monopod—and thus requires us to consider and examine the social implications of so-called sustainable development.

Only recently have scholars undertaken "cross-domain" analyses, where issues of equity, environment, and economy are considered. The concept of "just sustainability," which was put forth by Agyeman, Bullard, and Evans (2003), provides the conceptual foundation for exploring such joined-up thinking. In their account of sustainability Agyeman et al. fuse the environmental justice literature with the sustainability literature to argue convincingly that sustainability thus far has been a project of environmentalists but has ignored broader social concerns, especially those concerned with justice and equity. This work effectively problematizes the rhetorical space of sustainability and provides a language for local authorities, activists, and nongovernmental organizations (NGOs) to employ when advocating policies dedicated to just one leg of the sustainability stool. This work could thus be said to address Torgerson's (1995) comment that sustainability is so ambiguous that it allows actors from various backgrounds to proceed without agreeing on a single action. That is, before we make judgments about environmental sustainability, we must also consider other dimensions of sustainability. These accounts of sustainability, which range from outright sanguine to cautiously optimistic in tone, have come to define the bulk of the sustainability literature. They do, however, raise important conceptual questions about the ongoing project of sustainability, both as a catalyst for social change and a policy mechanism in urban and local contexts.

Critical commentaries along these lines have begun to emerge. Debates rage on the capacity of local authorities to act in the service of sustainability (Korten, 1996; Dryzek, 1997; Marvin & Guy, 1997; Lake, 2000; Gibbs, 2002) and on the meaning and usefulness of the concept itself (Myerson & Rydin, 1994; Torgerson, 1995; Berke & Manta-Conroy, 2001; Krueger & Agyeman, 2005). But, as Agyeman et al. (2003) point out, the debate is much broader than previously recognized. Lake's (2000) remarks on local sustainability initiatives in the United States support this. For Lake, many of the environmental benefits of urban sustainability initiatives accumulate in middle-class communities, often at the expense of peripheral local sites as well as those far away. Others have commented on the discursive nature of sustainability. Both Hajer (1995) and Harvey (1996) argue that sustainable development is a central story line for policy discourses such as ecological modernization. This notion of a "story line" suggests that sustainable development can be read as a new power/knowledge discourse for organizations seeking to accumulate power (Luke, 1995). This point is emphasized by Drummond and Marsden (1995), who conclude that the values and institutions of sustainability are those that prioritize the value of capital and the maintenance of existing patterns of social relations. Keil and Desfor (2003) suggest that sustainability discourses offer a counterhegemonic opportunity for subaltern groups to reshape urban environments, making them more equitable for disadvantaged groups. Krueger and Savage (2007) argue that those actors who focus upon social reproduction can deploy the discourse of sustainability—rather than discourses solely focused on environmental concerns—to produce outcomes that have a "red–green" tinge.

Returning to our point made at the outset of this chapter, "sustainability" is happening in a variety of concrete forms. There are more bike paths, policies for the protection of open space, local environmental improvement strategies, carbon reduction strategies, improved construction standards, and increased use of renewable energy sources, for example. The "story lines" of "sustainability" are being written and institutionalized across the board. As "sustainability" approaches critical mass as a development strategy in communities and nations around the world, it requires more detailed scrutiny. Analysis must come not just from those commentators we have mentioned but also from those with a keen understanding of the process and problems of social change in the context of capitalism's various forms, for the notion of sustainability is not ontologically fixed—nor are the outcomes of its implementation even across space or among social groups. Concepts of nature, scale, economic change, institutions, and governance must accompany sustainability analyses. Without these conceptual linkages we risk writing the story of sustainability in a way that merely fulfills the requirements of capitalist accumulation and thus rendering it something less than it is—a progressive project that ameliorates the negative externalities of economic activity for everyone, not just those characters who write the story.

THE STRUCTURE OF THIS VOLUME

The transfer of theoretical knowledge to practice is always a problematic one. However, the contributors to this book undertake this task elegantly by examining the sustainability debate from a variety of perspectives that combine the conceptual with the empirical. In the earliest chapters the contributors argue that we are experiencing a crisis in our relations with nature and that we need to shift toward a different set of relations through greater political engagement. As we have already indicated, Erik Swyngedouw (in Chapter 1) proposes that although virtually everyone is now in favor of sustainability, we are not "embarking on a different socioenvironmental trajectory." He argues that this inactivity is a product of the postpolitical condition, in particular the consensus around the inevitability of neoliberal capitalism as an economic system and a view that there is "one nature" that we can return to, given an appropriate set of sustainable development policies. Conversely, he suggests the key questions we need to ask are "What kinds of socioenvironmental arrangements do we wish to produce, how can this be achieved, and what sorts of natures do we wish to inhabit?" Without answers to such fundamentally political questions and the development of appropriate story lines that can be mobilized to address these, we effectively forestall any discussion of alternative socioeconomic futures and trajectories outside the neoliberal consensus. In this view, sustainability is therefore first and foremost a set of political questions about who benefits, who gains, and who loses from sustainable development. In a similar fashion, Roger Keil (Chapter 2) argues that the predominant contemporary form of sustainability is a particular interpretation of ecological modernization that redefines sustainable development as one possible route for the renewal of the capital accumulation process. Echoing Swyngedouw's arguments, he suggests that, despite widespread agreement on the "problematique" we face, any alternative "resolutique" that opposes the neoliberal consensus is not on the political agenda. As one counter to this, he proposes that we need to develop a radical urban political ecology that encourages alternatives to capitalism and that redirects the force of exchange value-oriented accumulation into use value-oriented products and services. Sue Buckingham continues with this analysis into the appropriate sites and forms of intervention in her chapter (Chapter 3) on the politics of gender in the theory and practice of sustainability. She examines those areas where existing policy and practice can be subject to "microruptures" to make space for changes whereby, in her example, environmental issues and gender inequality can be addressed in mutually constructive ways. Buckingham contends that women have the greatest traction for such microruptures at the smaller scale level of "microgeographies" (e.g., the household or neighborhood), as opposed to the more conventional focus upon the local and the urban in sustainability research.

The second set of chapters explores the contemporary politics and policies of sustainability at precisely this local and city–region scale. In particular, the authors illustrate contemporary debates about the intersection between sustainability, city–region competitiveness, and contemporary urban politics. In so doing, they develop the arguments contained in the first set of chapters through empirically grounded investigations of the tensions between the dominant neoliberal agenda and sustainability. They suggest that the fluid meanings of sustainability have enabled different groups of actors to redefine and manipulate the term in ways that suit their own political ends. Thus, David Gibbs and Rob Krueger (Chapter 4) draw upon empirical studies of the two new U.S. economy spaces of Austin and Boston to explore the conceptual linkages between sustainability and the new economy in order to shed some light on how sustainability might be congruent (or not) with neoliberal capitalist forms. Their findings indicate that in new economy spaces, economic success, quality of life, and a "good environment" are closely inter-twined. The types of sustainable development institutions and policy proscriptions that have emerged, however, are partially a result of the constraints of a broader neoliberal discourse. Both the chapters by Andy Jonas and Aidan White (Chapter 5) and Marc Parés and David Saurí (Chapter 6) draw upon an archetypal competitive entrepreneurial city—Barcelona—to explore the social dimensions of sustainability and possi-ble alternatives to neoliberal urbanism. In concurrence with Gibbs and Krueger, they conclude that the best cities "strategically select" particu-lar visions of urban sustainability aimed at livability rather than more radical solutions. Indeed, none of these authors suggests that competi-tiveness and sustainability have converged, but rather that actors in such spaces have adopted the sustainable development discourse and formu-lated policy options to address the tensions implicit in models of growth and economic competition, often involving new cross-class alliances and territorial coalitions. Similarly, the chapter by Anna Batchelor and Alan Patterson (Chapter 7) and that by Mike Raco (Chapter 8) both draw on case studies of local and regional government in South East England to illustrate the point that sustainable development initiatives have been constrained at the local scale by the strong orientation towards neo-liberal solutions by central government. For Batchelor and Patterson, al-though the region has become a significant administrative tier in the United Kingdom for sustainable development policy, a lack of clear lead-ership and deliberate "institutional muddle" both inhibit what can be achieved. Raco's examination of the Sustainable Communities Plan in England suggests that it has moved away from being something of a hy-brid between neoliberalism and sustainability toward trying to create a new spatial fix whereby the needs of accumulation are supported by the creation of new (built) environments in southeast England. In both these chapters the authors point to the disconnect between central government rhetoric and the resources and institutional powers needed to implement the policies. As Raco asks, "Can weak, 'hollowed-out' nation-states im-plement effective new regionalist agendas if those agendas require strong forms of state intervention and significant investment?"

The final two chapters focus upon the spatial politics involved in conservation planning and resource use. Echoing Swyngedouw's initial questions, Stephanie Pincetl and Basil Katz (Chapter 10) make the case that we need context-specific studies of sustainability that bring into question what is being sustained, at what scale, by, for whom, and which institutional mechanisms are being used. James Evans (Chapter 9) focuses upon the dominant fragmentation model of conservation planning, which he argues has been open to attack because of its division between "natural," referent landscapes and "human-influenced landscapes." As in Swyngedouw's and Keil's chapters, Evans points out that these ideas of "one nature" are misconceived and, indeed, are now largely absent among landscape ecologists. Despite this, he argues that fragmentation models remain dominant in conservation planning, because it "speaks the language of institutionalized power." Pincetl and Katz similarly focus upon power relations through a highly detailed study of water politics in Imperial Valley, California, as a way of addressing their set of key questions and to explore what they regard as the relative neglect of regulatory contexts in sustainability research. Pincetl and Katz argue that new water allocation rules in the valley have engendered a new "hydrosocial contract" between users and their environment and give a new exchange value to the water resource. A historic shift in water allocation away from agricultural use toward urban areas has been bound up with shifting structures, norms, and values that define the rights, constraints, and powers at work within the region. They conclude that, in order to understand how sustainable regulatory regimes can come about, we need much greater understanding of the organizational and institutional specificities involved over time and across space.

Campbell, who was referenced above, may be partly right: the details are crucial, but the process of translating "big public ideas" into "practice" is a messy, highly politicized one that cannot go unattended in the sustainability literature any longer. To be sure, "sustainability" actually exists as a diverse set of policy provisions being rolled out around the world, especially in western Europe and the United States, extending from international to local scales. What is unclear, however, is how those policies mesh with the social relations that attend our current form of capitalism and raise critical questions about the prospects of sustainability and how it must be engaged if it is to live up to its tripartite concerns of economic stability, social equity, and environmental integrity. It is the goal of this book's contributors to develop lines of conceptual analysis capable of capturing these intersecting phenomena: a multiscalar and globally reaching interest in sustainability in the context of contemporary capitalist social relations.

REFERENCES

Agyeman, J., Bullard, R., & Evans, B. (Eds.). (2003). *Just sustainabilities: Development in an unequal world*. Cambridge, MA: MIT Press.

Beatley, T. (2000). *Green urbanism: Learning from European cities*. Washington, DC: Island Press.

Beatley, T., & Manning, K. (1997). *The ecology of place: Planning for the environment, economy and community*. Washington, DC: Island Press.

Berke, P., & Manta-Conroy, M. (2000). Are we planning for sustainable development? *Journal of the American Planning Association, 66*(1), 21–33.

Brown, L. (2001). *Eco-economy: Building an economy for the Earth*. Washington, DC: Earth Policy Institute.

Calthorpe, P., & Fulton, W. (2001). *The regional city: Planning for the end of sprawl*. Washington, DC: Island Press.

Campbell, S. (1996). Green cities, growing cities, just cities: Urban planning and the contradictions of sustainable development. *Journal of the American Planning Association, 62*(3), 296–312.

Carley, M., & Spapens, P. (1998). *Sharing the world: Sustainable living and the global equity in the 21st century*. London: Earthscan.

Costanza, R., d'Arge, R., de Groot, R., Faber, S., Grasso, M., Hannon, B., et al. (1997). The value of the world's ecosystem services and natural capital. *Nature, 387*, 256–262.

Daly, H. (1974). The economics of the steady state. *American Economic Review, 64*, 15–21.

Drummond, I., & Marsden, T. K. (1995). Regulating sustainable development. *Global Environmental Change, 5*, 51–64

Dryzek, J. (1997). *The Politics of the Earth: Environmental discourses*. Oxford, UK: Oxford University Press.

Gearin, E. (2004). Smart growth or smart growth machine?: The smart growth movement and its implications. In J. Wolch, M. Pastor, & P. Drier (Eds.), *Up against sprawl: Public policy and the making of Southern California* (pp. 225–251). Minneapolis: University of Minnesota Press.

Gibbs, D. (2002). *Local development and the environment*. London: Routledge.

Hajer, M. (1995). *The politics of environmental discourse: Ecological modernisation and the policy process*. Oxford, UK: Oxford University Press.

Harvey, D. (1996). *Justice, nature and the geography of difference*. Oxford, UK: Blackwell.

Hawken, P., Lovins, A., & Lovins, H. (2000). *Natural capitalism: Creating the next industrial revolution*. Boston: Little, Brown.

International Council for Local Environmental Initiatives. (2002). *Second Local Agenda 21 survey*. New York: United Nations Commission on Sustainable Development.

Jacobs, M. (1999). Sustainable development as a contested concept. In A. Dobson (Ed.), *Fairness and futurity: Essays on environmental sustainability and social justice*. (pp. 21–45). Oxford, UK: Oxford University Press.

Keil, R., & Desfor, G. (2003). Ecological modernization in Los Angeles and Toronto. *Local Environment, 8*, 27–44.

Korten, D. (1996). Civic engagement in creating future cities. *Environment and Urbanization, 8*(1), 35–49.

Krueger, R., & Agyeman, J. (2005). Sustainability schizophrenia or "actually existing sustainabilities"?: Toward a broader understanding of the politics and promise of urban sustainability in the U.S. *Geoforum, 34*, 410–417.

Krueger, R., & Savage, L. (2007). City regions and social reproduction: A 'place' for sustainable development. *International Journal of Urban and Regional Research, 31*(1), 215–223.

Lake, R. (2000). Contradictions at the local state: Local implementation of the U.S. sustainability agenda in the USA. In N. Low, B. Gleeson, I. Elander, & R. Lidskog (Eds.), *Consuming cities: The urban environment in the global economy after the Rio Declaration* (pp. 70–90). London: Routledge.

Luke, T. (1995) Sustainable development as a power/knowledge system: The problem of "governmentality". In F. Fischer & M. Black (Eds.), *Greening environmental policy: The politics of a sustainable future* (pp. 21–32). London: Paul Chapman.

Marvin, S., & Guy, S. (1997). Creating myths rather than sustainability: The transition fallacies of the new localism. *Local Environment, 2*, 311–318.

Molotch, H. (1976). The city as a growth machine: Toward a political economy of place. *American Journal of Sociology, 82*(7), 309–332.

Myerson, G., & Rydin, Y. (1994). "Environment" and "planning": A tale of the mundane and the sublime. *Environment and Planning D: Society and Space, 12*, 437–452.

Pierce, J., & Dale, A. (1999). *Communities, development and sustainability across Canada.* Vancouver: UBC Press.

Portney, K. (2003). *Taking sustainable cities seriously.* Cambridge, MA: MIT Press.

Selman, P. (1996). *Local sustainability: Managing and planning ecologically sound places.* New York: St. Martin's Press.

Torgerson, D. (1995). The uncertain quest for sustainability: Public discourse and the politics of environmentalism. In F. Fischer & M. Black (Eds.), *Greening environmental policy: The politics of a sustainable future* (pp. 3–20). London: Paul Chapman.

Wackernagel, M., & Rees, W. (1996). *Our ecological footprint: Reducing human impact on the Earth.* Gabriola Island, BC, and New Haven, CT: New Society.

CHAPTER 1

Impossible "Sustainability" and the Postpolitical Condition

Erik Swyngedouw

Well, my dear Adeimantus, what is the nature of tyranny? It's obvious, I suppose, that it arises out of democracy.

—Plato, *The Republic*

Barbarism or Socialism.
—Karl Marx

Kyoto or the Apocalypse.
—Green saying

This chapter seeks to destabilize some of the most persistent myths about nature, sustainability, and environmental politics. First, I argue that there is no such thing as a singular Nature around which a policy of "sustainability" can be constructed. Rather, there are a multitude of natures and a multitude of existing or possible socionatural relations. Second, the obsession with a singular Nature that requires "sustaining" is fostered by an apocalyptic imaginary that forecloses asking serious political questions about possible socioenvironmental trajectories, particularly in the context of a neoliberal hegemony. Third, and most important, I argue that environmental issues and their political "framing" contribute to the making and consolidation of a postpolitical and postdemocratic condition, one that actually forecloses the possibility of a real politics of the environment. I conclude with a call to politicize the environment, one predicated upon the recognition of radically different possible socioenvironmental futures and the proliferation of new socioenvironmental imaginaries.

THE QUESTION OF NATURES

> Nature does not exist . . . or . . . When vegetarians will eat meat!

The *Guardian International* reported recently (August 13, 2005) how a University of Maryland scientist had succeeded in producing "cultured meat" (p. 1). Soon, he said, "it will be possible to substitute reared beef or chicken with artificially grown meat tissue. It will not be any longer necessary to kill an animal in order to get access to its meat. We can just rear it in industrialised labs." This might seem to be a magical solution that could tempt vegetarians to return to the flock of animal protein devotees while promising yet again (after the failed earlier promises made by the pundits of pesticides, the green revolution, and now genetic engineering and GM products) the final solution for world hunger and a more sustainable life for the millions of people who go hungry now. Meanwhile, NASA is spending about $40 million a year on how to recycle wastewater and return it to potable conditions, something that would of course be necessary to permit space missions of long duration—but which would be of significant importance on earth as well. At the same time, sophisticated new technologies are developed for sustainable water harvesting, for a more rational use of water, or for a better recycling of residual waters, efforts defended on the basis of the need to reach the Millennium Development Goals that promise, among other objectives, a reduction by half of the 2.5 billion people who do not have adequate access to safe water and sanitation.

In the meantime, other "natures" keep wreaking havoc around the world. The tsunami disaster of December 2004 comes readily to mind, as do the endless forest fires that blazed through Spain in the summer of 2005 during that country's driest summer on record, killing dozens of people and scorching the land; HIV continues its genocidal march through sub-Saharan Africa, and summer heat waves killed thousands of people prematurely in 2004 in France. In 2006 Europeans watched anxiously the nomadic wanderings of the avian flu virus, and all wait, almost stoically, for the moment it will pass more easily from birds to humans. While all this was going on, South Korea's leading biotech scientist, Hwang Woo Suk, proudly presented in August 2005 the Seoul National University Puppy ("Snuppy") to the global press as the first cloned dog (an Afghan hound), while a few months later, in December 2005, this science hero was forced to withdraw a paper on human stem cells from *Science* after accusations of intellectual fraud (later confirmed, prompting his resignation and exploding South Korea's great biotech dream). In the United Kingdom, male life expectancy between the "best" and "worst" areas is now more than 11 years, and the gap is widening, with life expectancy actually falling (for the first time since the Second World War) in some areas.[1] Tuberculosis is endemic again in East London; obesity is rapidly becoming the most seriously lethal socioecological condition in our fat cities (Marvin & Medd, 2006); and, as the ultimate cynical observation, nuclear energy is again celebrated and iconized by many elites, among them British Prime Minister Tony Blair, as the world's savior, the "enlightened" response to the climatic calamities promised by continuing carbon accumulation in our atmosphere while satisfying our seemingly insatiable need for energy.

This great variety of examples all testify to the blurring of boundaries between the human and the artificial, the technological and the natural, the nonhuman and the cyborg-human; they certainly also suggest that there are all manner of "natures" out there. While some of the above examples promise "sustainable" forms of development, others seem to stray further away from what might be labeled as sustainable. At first glance, "Frankenstein" meat, recycled water, and stem cell research are exemplary cases of possibly "sustainable" ways of dealing with apparently important socioenvironmental problems while solving significant social problems (animal ethics and food supply on the one hand, dwindling freshwater resources or unsustainable body metabolisms on the other). Sustainable processes are sought for around the world, and solutions for our precarious environmental condition are feverishly developed. Sustainability, so it seems, is in the making, even for vegetarians.

Meanwhile, as some of the other examples attest, socioenvironmental processes keep on wreaking havoc in many places around the world. "Responsible" scientists, environmentalists of a variety of ideological stripes and colors, together with a growing number of world leaders and politicians, keep on spreading apocalyptic and dystopian messages about the clear and present danger of pending environmental catastrophes that will be unleashed if we refrain from immediate and determined action. Particularly the threat of global warming is framed in apocalyptic terms if the atmospheric accumulation of CO_2 (which is of course the classic "side effect" of the accumulation of capital in the troposphere) continues unabated. Table 1.1 recounts some of the most graphic recent media headlines on the doomsday theme. The world as we know it will come to a premature end (or be seriously mangled) unless we urgently reverse, stop, or at least slow down global warming and return the climate to its status quo ante. Political and regulatory mechanisms (such as the Kyoto Protocol) and CO_2 reducing technomachinery (such as hybrid cars) are developed that would—so the hope goes—stop the threatening evolution and return the earth's temperature to its benevolent earlier condition. From this perspective, sustainability is predicated upon a return, if possible, to a perceived global climatological equilibrium that would permit sustainable continuation of the world's current way of life.

So, while one sort of sustainability seems to be predicated upon feverishly developing new natures (like artificial meat, cloned stem cells, or manufactured clean water), forcing nature to act in a way we deem sustainable or socially necessary, the other type is predicated upon limiting or redressing our intervention in nature, returning it to a presumably more benign condition so that human and non-human sustainability in the medium and long term can be assured. Despite the apparent contra-

TABLE 1.1. Warnings of the Apocalypse: Recent Media Headlines

- "Global Warming and Ozone Loss: Apocalypse Soon"[a]
- "Sea Levels Likely to Rise Much Faster Than Was Predicted"[a]
- "Global Warming Is Causing the Greenland Ice Cap to Disintegrate Far Faster Than Anyone Predicted"[a]
- "Global Warming 30 Times Quicker Than It Used to Be"[a]
- "Climate Change: On the Edge"[a]
- "Water Wars"[b]
- "The Four Horsemen of Industrial Society: War, Over-Population, Climate Change & Peak Oil"[c]
- "Pentagon Warns Bush of Apocalyptic Climate Change by 2020"[d]

[a]From *The Independent*, February 17, 2006.
[b]From *The Independent*, February 28, 2006.
[c]From *Energy Bulletin*, January 12, 2006.
[d]From *The Observer*, February 22, 2004.

dictions of these two ways of "becoming sustainable" (one predicated upon preserving nature's status quo, the other predicated upon producing new natures), they share the same basic vision that technonatural and sociometabolic interventions are urgently needed if we wish to secure the survival of the planet and much of what it contains. But these examples also show that "nature" is not always what it seems to be. Frankenstein meat, dirty water, bird–flu virus symbiosis, stem cells, fat bodies, heat waves, tsunamis, hurricanes, genetic diversity, CO_2, to name just a few, are radically different things, expressing radically different natures, pushing in radically different directions, with radically different consequences and outcomes, and with radically different human/ nonhuman connectivities. If anything, before we can even begin to unpack "sustainability," these examples certainly suggest that we urgently need to interpolate our understanding of nature and revisit what we mean by nature and what we assume nature to be.

Surrendering Nature: Indeterminate Natures

Slavoj Žižek suggests in *Looking Awry* that the current ecological crisis is indeed a radical condition that not only constitutes a real and present danger, but, equally importantly, "questions our most unquestionable presuppositions, the very horizon of our meaning, our everyday understanding of 'nature' as a regular, rhythmic process" (Žižek, 1992/2002a: 34). It raises serious questions about what were long considered self-evident certainties. He argues that this fundamental threat to our deepest convictions of what we always thought we knew for certain about nature is co-constitutive of our general unwillingness to take the ecological crisis completely seriously. It is this destabilizing effect that explains "the fact that the typical, predominant reaction to it still consists in a variation of the famous disavowal, 'I know very well (that things are deadly serious, that what is at stake is our very survival), but just the same I don't really believe, . . . and that is why I continue to act as if ecology is of no lasting consequence for my everyday life' " (Žižek, 2002: 35). The same unwillingness to question our very assumptions about what nature is (and even more so what natures might "become") also leads to the typical obsessive reactions of those who *do* take the ecological crisis seriously. Žižek considers both the case of the environmental activist, who in his or her relentless and obsessive activism to achieve a transformation of society in more ecologically sustainable ways, expresses a fear that to stop acting would lead to catastrophic consequences. In his words, obsessive acting becomes a tactic to stave off the ultimate catastrophe, that is, "If I stop doing what I am doing, the world will come to an end in an ecological Armageddon." Other observers, of course, see all manner of transcendental signs in the "revenge of nature," read it as a message that signals our destructive intervention in nature, and urge us to change our relationship with nature.

In other words, we have to listen to nature's call, as expressed by the pending environmental catastrophe, and respond to its message that pleads for a more benign associational relationship with nature, a posthuman affective connectivity, as a cosmopolitical "partner in dialogue." While the first attitude radically ignores the reality of possible ecological disaster, the second, which is usually associated with actors defending "sustainable" solutions for our current predicament, is equally problematic in that it ignores, or is blind to, the inseparable gap between our symbolic representation (our understanding) of Nature and the actual acting-out of a wide range of radically different, and often contingent, natures. In other words, there is—of necessity—an unbridgeable gap, a void, between our dominant view of Nature (as a predictable and determined set of processes that tends toward a (dynamic) equilibrium—but one that is disturbed by our human actions and that can be "rectified" with proper sustainable practices—and the acting-out of natures as an (often) unpredictable, differentiated, incoherent, open-ended, complex, chaotic (although by no means unordered or unpatterned) set of processes. The latter implies the existence not only of many natures, but, more importantly, it also assumes the possibility of all sorts of possible future natures, all manner of imaginable different human–nonhuman assemblages and articulations, and all kinds of different possible socioenvironmental becomings.

The inability to take "natures" seriously is dramatically illustrated by the controversy over the degree to which disturbing environmental change is actually taking place and the risks or dangers associated with it. Lomberg's *The Skeptical Environmentalist* (1998) captures one side of this controversy in all its phantasmagorical perversity, while climate change doomsday pundits represent the other. Both sides of the debate argue from an imaginary position of the presumed existence of a dynamic balance and equilibrium—the point of "good" nature—but one side claims that the world is veering off the correct path, while the other side (Lomborg and other sceptics) argues that we are still pretty much on nature's course. With our gaze firmly fixed on capturing an imaginary "idealized" Nature, the controversy further solidifies our conviction of the possibility of a harmonious, balanced, and fundamentally benign *One Nature* if we would just get our interaction with it right, an argument blindly (and stubbornly) fixed on the question of where Nature's rightful point of benign existence resides. This futile debate, circling around an assumedly centered, known, and singular Nature, certainly permits—in fact, invites—imagining ecological catastrophe at some distant point (global burning, or freezing, through climate change, resource depletion, death by overpopulation). Indeed, imagining catastrophe and fantasizing about the final ecological Armageddon seems considerably easier for most environmentalists than envisaging relatively small changes in the sociopolitical and cultural-economic organization of local and global life here and now. Or, put differently, the world's premature ending in a climatic Armageddon seems easier to imagine (and sell to the public) than a transformation of (or end to) the neoliberal capitalist order that keeps on practicing expanding energy use and widening and deepening its ecological footprint.

These sorts of considerations led Slavoj Žižek controversially to state that "nature does not exist." Of course, he does not imply that there are no such "things" as quarks or other subatomic particles, black holes, tsunamis, sunshine, trees, or HIV viruses. Even less would he deny the radical effects of CO_2 and other greenhouse gases on the climate or the lethal consequences of water contamination for the world's poor. On the contrary, they are very real, many posing serious environmental problems, occasionally threatening entire populations (e.g., AIDS), but he insists that the "Nature" we see and work with is necessarily radically imagined, scripted, symbolically charged, and radically distant from the various natures that are out there, which are complex, chaotic, often unpredictable, often radically contingent, risky, patterned in endlessly complex ways, and ordered along "strange" attractors. In other words, there is no balanced, dynamic equilibrium-based nature out there that needs or requires salvation in the name of either "Nature" itself or of an equally imagined universal human survival. "Nature" simply does not exist. There is nothing foundational in nature that needs, demands, or requires sustaining. The debate and controversies over nature and what to do with it, in contrast, signal our political inability to engage in directly political and social argument and strategies about rearranging the social coordinates of everyday life and the arrangements of sociometabolic organization (something usually called capitalism) that we inhabit. In order words, imagining a benign and "sustainable" Nature avoids asking the politically sensitive but vital question as to what kind of socioenvironmental arrangements we wish to produce, how these can be achieved, and what sort of natures we wish to inhabit.

WHAT IS ENVIRONMENT?: IMPOSSIBLE SUSTAINABILITY OR UNDESIRABLE SUSTAINABILITY?

Desiring (Un)sustainability

So, if "Nature" does not exist, what, then, can one say about sustainability, a concept (and associated set of fuzzy practices) that is deeply indebted and intensely connected to the particular imaginings of Nature suggested above. Since its early definition in the Brundtland Commission report of 1987, the concept (but not much of the practice) of "sustainability" has really taken off. A cursory glance at both popular and academic publications will quickly suggest a whole array of "sustainabilities": sustainable environments, sustainable development, sustainable growth, sustainable wetlands, sustainable bodies, sustainable companies, sustainable processes, sustainable incomes, sustainable cities, sustainable technologies, sustainable water provision, even sustainable poverty, sustainable accumulation, sustainable markets, and sustainable loss. I have not been able to find a single source that is against "sustainability." Greenpeace is in favor, George Bush Jr. and Sr. are, the World Bank and its chairman (a prime warmonger on Iraq) are, the Pope is, my son Arno is, the rubber tappers in the Brazilian Amazon forest are, Bill Gates is, the labor unions are. All are presumably concerned about the long-term socioenvironmental survival of (parts of) humanity; most just keep on doing business as usual. The clear and present danger posed by the environmental question is obviously not dramatic enough to be taken seriously in terms of embarking on a different socioenvironmental trajectory. That is left to do at some other time—and certainly not before the day after tomorrow. Of course, this cacophony of voices and imaginations also points to the inability to agree on the meaning of, or, better, to the lack of, a singular "Nature." There are obviously multiple imaginations that mobilize or appropriate sustainability as radically and truthfully theirs, based on equally imaginative variations of what constitutes "Nature."

Environmentalists (whether activists or scientists) invariably invoke the global physical processes that threaten our existence and insist on the need to re-engineer nature so that it can return to a "sustainable" path. Armed with their charts, formulas, models, numbers, and grant applications, to which activists usually add the inevitable pictures of scorched land, factories or cars emitting carbon fumes, dying animals and plants, suffering humans, apocalyptic rhetoric, and calls for subsidies and financial support, scientists, activists, and all manner of assorted other human and nonhuman actants enter the domain of the social, the public, and, most importantly, the political. Thus "natures" enter the political realm. A particular and symbolically enshrined nature enters the parliament of politics, but does so in a duplicitous manner. It is a treacherously deceitful Nature that enters politics, one that is packaged, numbered, calculated, coded, modeled, and represented by those who claim to possess, know, understand, and speak for the "real Nature." In other words, what enters the domain of politics are the coded and symbolized versions of nature mobilized by scientists, activists, industrialists, and the like. This is particularly evident in examples such as the debate over genetically modified organisms (GMOs), global climate change, BSE ("mad cow disease"), biodiversity loss, and other equally pressing issues.

Invariably, the acting-out of Nature—as scripted by the bearers of nature's knowledge—enters the political machinery as coded language that also already posits its political and social solution and does not tolerate, in the name of Nature, dissent other than that framed by its own formulations. It is in this sense, of course, that the argument about climate change is exclusively formulated in terms of believers and nonbelievers, as a quasi-religious faith, but the weapons of the struggle in this case are matters of fact such as data, models, and physicochemical analyses. And the solutions to the question of sustainability are already prefigured by the way in which Nature is made to speak. Creeping increases in long-term global temperatures, which will cause untold suffering and damage, are caused by CO_2 output. Hence, the solution to future climate ills resides in cutting back on CO_2 emissions. Notwithstanding the validity of the role of CO_2 in co-constituting the process of climate change, the problematic of the future calamities the world faces is posited primarily in terms of the physical acting-out of one of nature's components, CO_2, as is its solution found in bringing CO_2 within our symbolic (socioeconomic) order, futilely attempted with the Kyoto agreement or other neoliberal market-based mechanisms. Questioning the politics of climate change in itself is already seen as an act of treachery, as an unlawful activity, banned by "Nature" itself.

Undesirable Sustainability: Environmental Politics as Postpolitics

Although there may be no all-embracing singular conception of "Nature," there certainly is a politics of nature or a politics of the environment. The collages of apparently contradictory and overlapping vignettes of the environmental conditions outlined above share one common threat that many of us, Bush and Blair, my son and Greenpeace, Oxfam and the World Bank, agree on. The world is in environmental trouble. And we need to act politically now.

Both the December 2004 tsunami and New Orleans's Katrina brought the politicization of nature home with a vengeance. Although the tsunami had everything to do with the earth's geodetic acting-out and with the powerless of South East Asia drowning in its consequences and absolutely nothing to do with climate change or other environmentally degrading practices, the tsunami calamity was and continues to be staged as a socioenvironmental catastrophe, another assertion of the urgent need to revert to more "sustainable" socioenvironmental practices.

New Orleans's socioenvironmental disaster was of a different kind. While there may be a connection between the number and intensity of hurricanes and climate change, that of course accounts for neither the dramatic destruction of poor people's lives in the city nor for the plainly blatant racist spectacles that were fed into the media on a daily basis in the aftermath of the hurricane's rampage through the city. The imaginary staged in the aftermath of the socioenvironmental catastrophe of New Orleans singled out disempowered African Americans twice, first as victims, then as criminals. Even *The New York Times* conceded that 80% of the reported "crimes" taking place in unruly and disintegrating New Orleans in the aftermath of the hurricane's devastation were based on rumor and hearsay—a perverse example of how liberal humanitarian concern may be saturated with racialized coding and moralizing against the poorest and most excluded portions of society. Of course, after the poor were hurricaned out of New Orleans, the wrecked city is rapidly turning into a fairy tale playground for urban developers and city boosters who will make sure, this time around, that New Orleans will be rebuilt in their image of a sustainable capitalist city: green, white, rich, conservative, and neoliberal (Davis, 2006).

The popular response to Katrina, the barrage of apocalyptic warnings of the pending catastrophes wreaked by climate change, and the need to take urgent remedial action to engineer a retrofitted "balanced" climate are perfect examples of the tactics and configurations associated with the present postpolitical condition, primarily in the United States and Europe. Indeed, a politics of sustainability, predicated upon a radically conservative and reactionary view of a singular—and ontologically stable and harmonious—Nature is necessarily one that eradicates or evacuates the "political" from debates over what to do with natures. The key political question is one that centers on the question of what kinds of natures we wish to inhabit, what kinds of natures we wish to preserve, to make, or, if need be, to wipe off the surface of the planet (e.g., the HIV virus), and on how to get there.

The fantasy of "sustainability" imagines the possibility of an originally fundamentally harmonious Nature, one that is now out-of-synch but which, if "properly" managed, we can and have to return to by means of a series of technological, managerial, and organizational fixes. As suggested above, many, from different social, cultural, and philosophical position, agree with this dictum. Disagreement is allowed, but only with respect to the choice of technologies, the mix of organizational fixes, the details of the managerial adjustments, and the urgency of the timing and implementation. Nature's apocalyptic future, if unheeded, symbolizes and nurtures the solidification of the postpolitical condition. And the excavation and critical assessment of this postpolitical condition, nurtured and embodied by most of current Western socioenvironmental politics, is what we shall turn to next.

THE POSTPOLITICAL POSTDEMOCRATIC CONDITION: EVACUATING SOCIOENVIRONMENTAL POLITICS

Postpolitical!?

Slavoj Žižek and Chantal Mouffe, among others, define the postpolitical as a political formation that actually forecloses the political, that prevents the politicization of particulars (Žižek, 1999a: 35, 2006; Mouffe, 2005). A situation or an event becomes

> political when a particular demand (cutting greenhouse gases, stopping the exploitation of a particular resource and so on) starts to function as a metaphoric condensation of the global opposition against Them, those in power, so that the protest is no longer just about that demand, but about the universal dimension that resonates in that particular demand. . . . What post-politics tends to prevent is precisely this meta-

phoric universalization of particular demands: post-politics mobilizes
the vast apparatus of experts, social workers, and so on, to reduce the
overall demand (complaint) of a particular group to just this demand,
with its particular content—no wonder that this suffocating closure
gives birth to "irrational" outbursts of violence as the only way to give
expression to the dimension beyond particularity. (Žižek, 1999b: 204)

In Europe and the United States, in particular, such postpolitical arrangements are largely in place. Postpolitics rejects ideological divisions
and the explicit universalization of particular political demands. Instead,
the postpolitical condition is one in which a consensus has been built
around the inevitability of neoliberal capitalism as an economic system,
parliamentary democracy as the political ideal, and humanitarianism
and inclusive cosmopolitanism as a moral foundation. As Žižek (1999b:
198) puts it:

> In post-politics, the conflict of global ideological visions embodied in
> different parties which compete for power is replaced by the collabora-
> tion of enlightened technocrats (economists, public opinion special-
> ists, . . .) and liberal multiculturalists; via the process of negotiation of
> interests, a compromise is reached in the guise of a more or less universal
> consensus. Post-politics thus emphasizes the need to leave old ideologi-
> cal visions behind and confront new issues, armed with the necessary ex-
> pert knowledge and free deliberation that takes people's concrete needs
> and demands into account.

Postpolitics is thus about the administration of social or ecological
matters, and they remain of course fully within the realm of the possible,
of existing socioecological relations. "The ultimate sign of post-politics
in all Western countries," Žižek (2002b: 303) argues, "is the growth of a
managerial approach to government: government is reconceived as a
managerial function, deprived of its proper political dimension." Postpolitics discourages politicization in the classical Greek sense, that is, as
the metaphorical universalization of particular demands, which aims at
"more" than negotiation of interests:

> The political act (intervention) proper is not simply something that
> works well within the framework of existing relations, but something
> that *changes the very framework that determines how things work.* . . .
> [A]uthentic politics . . . is the art of the *impossible*—it changes the very
> parameters of what is considered "possible" in the existing constella-
> tion. (Žižek, 1999b: 199, emphasis in the original)

A genuine politics, therefore, is "the moment in which a particular demand is not simply part of the negotiation of interests but aims at something more, and starts to function as the metaphoric condensation of the global restructuring of the entire social space" (Žižek, 1999b: 208). Genuine politics is about the recognition of conflict as constitutive of the social condition, and the naming of the socio-ecological spaces that can become. The political becomes for Žižek and Rancière the space of litigation (Žižek, 1998), the space for those who are not-All, who are uncounted and unnamed, not part of the "police" (symbolic or state) order. A true political space is always a space of contestation for those who have no name or no place. As Diken and Laustsen (2004: 9) put it: "Politics in this sense is the ability to debate, question and renew the fundament on which political struggle unfolds, the ability to radically criticise a given order and to fight for a new and better one. In a nutshell, then, politics necessitates accepting conflict." A radical-progressive position "should insist on the unconditional primacy of the inherent antagonism as constitutive of the political" (Žižek, 1999a: 29).

Postpolitical parliamentary rule, in contrast, permits the politicization of everything and anything, but only in a noncommittal way and as nonconflict. Absolute and irreversible choices are kept away; politics becomes something one can do without making decisions that divide and separate (Thomson, 2003). A consensual postpolitics arises thusly, one that either eliminates fundamental conflict (i.e., we all agree that climate change is a real problem that requires urgent attention) or elevates it to antithetical ultrapolitics. Those who deny the realities of a dangerous climate change are blinded radicals that put themselves outside the legitimate social (symbolic) order. The same "fundamentalist" label is of course also put on those who argue that dealing with climate change requires a fundamental reorganization of the hegemonic neoliberal capitalist order. The consensual times we are currently living in have thus eliminated a genuine political space of disagreement. However, consensus does not equal peace or absence of fundamental conflict (Rancière, 2005a: 8). Under a postpolitical condition, "Everything is politicised, can be discussed, but only in a non-committal way and as a non-conflict. Absolute and irreversible choices are kept away; politics becomes something one can do without making decisions that divide and separate. When pluralism becomes an end in itself, real politics is pushed to other arenas" (Diken & Laustsen, 2004: 7), in the present case to street rebellion and protest, and terrorist tactics (e.g., the animal liberation movement in the United Kingdom).

Spivak might be in their party

Difficulties and problems, such as environmental concerns that are generally staged and accepted as problematic, need to be dealt with through compromise, managerial and technical arrangements, and the production of consensus. "Consensus means that whatever your personal commitments, interests and values may be, you perceive the same things, you give them the same name. But there is no contest on what appears, on what is given in a situation and as a situation" (Rancière, 2003a: §4). The key feature of consensus is "the annulment of dissensus . . . the 'end of politics' " (Rancière, 2003b: §32). The most utopian alternative to capitalism left at our disposal is to develop postpolitical alternatives to creating a more just and sustainable society, since it would not make any economic sense not to do so. Of course, this postpolitical world eludes choice and freedom (other than those alternatives tolerated by the consensus). And in the absence of real politicization of particulars, the only position of real dissent is that of either the traditionalists (those stuck in the past who refuse to accept the inevitability of the new global neoliberal order) or the fundamentalists. The only way to deal with them is by sheer violence, by suspending their "humanitarian" and "democratic" rights. The postpolitical relies on either including all in a consensual pluralist order or excluding radically those who posit themselves to be outside the consensus. For them, as Agamben (2005) argues, the law is suspended; they are literally put outside the law and treated as extremists and terrorists.

Not only are the environment and debates over the environment and nature perfect expressions of such a postpolitical order, but also, in fact, the mobilization of environmental issues is one of the key arenas through which this postpolitical consensus becomes constructed, when "politics proper is progressively replaced by expert social administration" (Žižek, 2005a: 117). The fact that Bush does not want to play ball on the climate change issue is indeed seen by both the political elites in Europe and the environmentalists as a serious threat to the postpolitical consensus. That is why both political elites and opposition groups label him as a radical conservative. U.S. president Bill Clinton, of course, embodied the postpolitical consensus in a much more sophisticated and articulated manner, not to speak of his unfortunate second-in-command, Al Gore, who recently resurfaced as a newborn climate change warrior (The Independent, May 22, 2006).

The postpolitical environmental consensus, therefore, is one that is radically reactionary, one that forestalls the articulation of divergent, conflicting, and alternative trajectories of future socioenvironmental possibilities and of human–human and human–nature articulations and assemblages. It holds on to a harmonious view of nature that can be recaptured while reproducing if not solidifying a liberal capitalist order for which there seems to be no alternative. Much of the sustainability argument has evacuated the politics of the possible, the radical contestation of alternative future socioenvironmental possibilities and socionatural arrangements, and attempts to silence the radical antagonisms and conflicts that are constitutive of our socionatural orders by externalizing conflict. In climate change, for example, the conflict is posed as one of society versus CO_2. In fact, the sustainable future desired by "sustainablity" pundits has no name. While alternative futures in the past were named and counted (for example, communism, socialism, anarchism, libertarianism, liberalism), the desired sustainable environmental future has no name and no process, only a state or condition. This is as exemplified by the following apocalyptic warning in which the celebrated quote from Marx's Communist Manifesto and its invocation of the "the specter of communism that is haunting the world" (once the celebrated name of hope for liberation) is replaced by the specter of Armageddon:

> A specter is haunting the entire world: but it is not that of communism. . . . Climate change—no more, no less than nature's payback for what we are doing to our precious planet—is day by day now revealing itself. Not only in a welter of devastating scientific data and analysis but in the repeated extreme weather conditions to which we are all, directly or indirectly, regular observers, and, increasingly, victims. (Levene, 2005)

Climate change is of course not a politics, let alone a political program or socioenvironmental project; it is pure negation of all that is political; a type of negation we can all concur with, around which a consensus can be built, but which eludes conflict and evacuates the political field. By doing so, it does not translate Marx's dictum to the contemporary period but rather turns it into its radical travesty.

From Postpolitics to Postdemocracy

There is of course a close relationship between the postpolitical condition and the functioning of the political system. In particular, the postpolitical threatens the very foundation upon which a democratic polity rests. Indeed. French philosopher Jacques Rancière defines this kind of consensual postpolitics as harboring a "postdemocracy," rather than seeing it as forming a deepened democracy:

> Postdemocracy is the government practice and conceptual legitimation of a democracy *after the demos*, a democracy that has eliminated the appearance, miscount, and dispute of the people and is thereby reducible to the sole interplay of state mechanisms and combinations. . . . It is the practice and theory of what is appropriate with no gap left between the forms of the State and the state of social relations. (Rancière, 1995: 142–153; also in Mouffe, 2005: 29)

In this postdemocratic postpolitical era, in contrast, adversarial politics (of the left/right variety or of radically divergent struggles over imagining socioenvironmental futures, for example) are considered hopelessly out of date. Although disagreement and debate are of course still possible, they operate within an overall model of consensus and agreement (Crouch, 2004). There is, indeed, in the domain of environmental policies and politics, a widespread consensus that Nature and the Environment need to be taken seriously and that appropriate managerial–technological apparatuses can and should be negotiated to avoid imminent environmental catastrophe. At the same time, of course, there is hegemonic consensus that no alternative to liberal global hegemony is possible.

This postpolitical framework is of course politically correlative to the theoretical argument advanced most coherently by sociologists Ulrich Beck or Anthony Giddens. They argue that adversarial politics organized around collective identities that were shaped by the internal relations of class-based capitalism are replaced by an increasingly individualized, fragmented, "reflexive" series of social conditions. For Beck, for example, "simple modernization ultimately situates the motor of social change in categories of instrumental rationality (reflection), 're-flexive' modernization conceptualizes the motive power of social change in categories of the side-effect (reflexivity). Things at first unseen and un-reflected, but externalized, add up to structural rupture that separates industrial from new modernities in the present and the future" (Beck, 1997: 38). From this perspective, "The distinction between danger (characteristic of pre-modern and modern societies) and risk (the central aspect of late modern risk society) refers to technological change. However, the transition from danger to risk can be related to the . . . process

of the weakening of the state. In risk society what is missing is an authority that can symbolise what goes wrong. Risk is, in other words, the
danger that cannot be symbolised" (Diken & Laustsen, 2004: 11; see
also Žižek, 1999b: 322–347), that has no name. Politicization, then, is to
make things enter the parliament of politics (see Latour, 2004), but the
postdemocratic condition does so in a consensual conversation in tune
with the postpolitical evacuation of real antagonism. The environmental
apocalypse in the making puts the state on the spot (e.g., with BSE, avian
flu, climate change), yet exposes the impotence of the state to "solve" or
"divert" the risk and undermines the citizens' sense of security guaran-
teed by the state.

It is these "side-effects" identified by Ulrich Beck (such as, for example, the accumulation of CO_2) that are becoming the key arenas
around which political configuration and action crystallize, and of
course (global) environmental problems are the classic example of such
effects, unwittingly produced by modernization itself but now requiring
second "reflexive" modernization to deal with. The old left/right collective politics that were allegedly generated from within the social relations that constituted modernity are no longer, if they ever were, valid or
performative. This, of course, also means that the traditional theaters of
politics (state, parliament, parties, etc.) are not any longer the exclusive
terrain of the political: "The political constellation of industrial society
is becoming unpolitical, while what was unpolitical in industrialism is
becoming politicals" (Beck, 1994: 18). It is exactly the side effects (the
risks) of modernizing globalization that need management, that require
politicization. A new form of politics (what Rancière, Žižek, and Mouffe
define exactly as postpolitics) thus arises, what Beck (1994: 22) calls
subpolitics: — same idea

> "Sub-politics" is distinguished from "politics" in that (a) agents outside
> the political or corporatist system are allowed also to appear on the
> stage of social design (this group includes professional and occupational
> groups, the technical intelligentsia in companies, research institutions
> and management, skilled workers, citizens' initiatives, the public sphere
> and so on), and (b) not only social and collective agents but individuals
> as well compete with the latter and each other for the emerging power
> to shape politics.

Chantal Mouffe (2005: 40–41) summarizes Beck's prophetic vision
of a new democracy as follows:

In a risk society, which has become aware of the possibility of an ecological crisis, a series of issues which were previously considered of a private character, such as those concerning the lifestyle and diet, have left the realm of the intimate and the private and have become politicized. The relation of the individual to nature is typical of this transformation since it is now inescapably interconnected with a multiplicity of global forces from which it is impossible to escape. Moreover, technological progress and scientific development in the field of medicine and genetic engineering are now forcing people to make decisions in the field of "body politics" hitherto unimaginable. . . . What is needed is the creation of forums where a consensus could be built between the experts, the politicians, the industrialists and citizens on ways of establishing possible forms of co-operation among them. This would require the transformation of expert systems into democratic public spheres.

This postpolitical constitution, which we have elsewhere defined as new forms of autocratic governance-beyond-the-state (Swyngedouw, 2005), reconfigures the act of governing to a stakeholder-based arrangement of governance in which the traditional state forms (national, regional, or local government) partake together with experts, NGOs, and other "responsible" partners (see Crouch, 2004). Not only is the political arena divested of radical dissent, critique, and fundamental conflict, but the parameters of democratic governing itself are being shifted, announcing new forms of governmentality, in which traditional disciplinary society is transfigured into a society of control through disembedded networks (like the Kyoto Protocol, the Dublin Statement, the Rio Summit, etc.). These new global forms of "governance" are expressive of the postpolitical configuration (Mouffe, 2005: 103):

> Governance entails an explicit reference to "mechanisms" or "organized and coordinated activities" appropriate to the solution of some specific problems. Unlike government, governance refers to "policies" rather than "politics" because it is not a binding decision-making structure. Its recipients are not 'the people' as collective political subject, but 'the population' that can be affected by global issues such as the environment, migration, or the use of natural resources. (Urbinati, 2003: 80)

Anthony Giddens (1991, 1994, 1998) has also been a key intellectual interlocutor of this postpolitical consensus. He argues that globalized modernity has brought in its wake all manner of uncertainties as a result of humans' proliferating interventions in nature and in social life, resulting in an explosive growth of all sorts of environmental and life-related issues. The ensuing "life politics is about the challenges that face collective humanity" (Giddens, 1994: 10). What is required now, in a context of greater uncertainty but also with enhanced individual autonomy to make choices, is to generate active "trust" achieved through a "dialogic democracy." Such "dialogic" mode is exactly the consensual politics Jacques Rancière defines as postdemocratic (Rancière, 1995, 2005b). As Chantal Mouffe (2005: 45) maintains, "Active trust implies a reflexive engagement of lay people with expert systems instead of their reliance on expert authority." Bruno Latour, in his politics of nature, of course equally calls for such a new truly democratic cosmopolitical constitution through which both human and nonhuman actants enter into a new public sphere, where matters of fact are turned into matters of concern, articulated, and brought together through heterogeneous and flat networks of related and relationally constituted human/nonhuman assemblages (Latour, 2004, 2005). Nothing is fixed, sure, or given, and everything is continuously in doubt, negotiated, or brought into the political field. Political space is not a contingent space where that which has no name is brought into the discussion, is given a name, and is counted, but rather things and people are "hailed" to become part of the consensual dialogue, of the dialogic community. The question remains, of course, of "Who does what sort of hailing?" Thinking about truth and falsity, doubt and certainty, right or wrong, friend or foe, would no longer be possible; the advent of a truly cosmopolitan order in a truly cosmopolitical constitution looms around the corner as the genuine possibility in the new modernity (Stengers, 2003).

In the domain of the environment, climate change, biodiversity preservation, sustainable sociotechnical environmental entanglements and the like exemplify the emergence of this new postpolitical configuration; they are an unexpected and unplanned by-product of modernization, they affect the way we do things, and, in turn, a new politics emerges to deal with them. This liberal cosmopolitical "inclusive" politics suggested by Beck and his fellow travelers as a radical answer to unbridled and unchecked neoliberal capitalist globalization, of course, is predicated upon three assumptions:

1. The social and ecological problems caused by modernity/capitalism are external side effects; they are not an inherent and integral part of the deterritorialized and reterritorialized relations of global neoliberal capitalism. That is why we speak of the ex-

cluded or the poor, and not about social power relations that produce wealth and poverty, or empowerment and disempowerment. A strictly populist politics emerges here, one that elevates the interest of "the people," "nature," or "the environment" to the level of the universal rather than aspiring to universalize the claims of particular natures, environments, or social groups or classes.

2. These side effects are posited as global, universal, and threatening: they are a total threat, of apocalyptic nightmarish proportions.

3. The "enemy" or the target of concern is thereby of course continuously externalized. The "enemy" is always vague, ambiguous, and ultimately vacant, empty, and unnamed (CO_2, gene pools, desertification, etc.). They can be managed through a consensual dialogical politics. Demands become depoliticized, or, rather, radical politics comes to be not about demands but about things.

ENVIRONMENTAL POPULISM VERSUS A POLITICS OF THE ENVIRONMENT

The postpolitical condition articulates, therefore, with a populist political tactic as the conduit, to instigate "desirable" change. Environmental politics and debate over "sustainable" futures in the face of pending environmental catastrophe are a prime expression of the populist ploy of the postpolitical postdemocratic condition. In this part, we shall chart the characteristics of populism (see, among others, Canovan, 1999; Laclau, 2005; Mouffe, 2005; Žižek, 2005b) and how this is reflected in mainstream environmental concerns.

First, populism invokes *The* Environment and *The* people (if not humanity as a whole) in a material and philosophical manner. All people are affected by environmental problems, and the whole of humanity (as well as large parts of the nonhuman) is under threat from environmental catastrophes. At the same time, the environment is running wild, veering off the path of (sustainable) control. As such, populism cuts across the idiosyncrasies of different human and nonhuman "natures" and their specific "acting-outs," silences ideological and other constitutive social differences, and papers over conflicts of interests by distilling a common threat or challenge to both nature and humanity.

Second, populism is based on a politics of "the people know best" (although "the people" often remains unspecified and unnamed), supported by an assumedly neutral scientific technocracy, and advocates a direct relationship between people and political participation. It is assumed that this will lead to a good, if not optimal, solution, a view strangely at odds with the presumed radical openness, uncertainty, and undecidability of the excessive risks associated with Beck's or Giddens's second modernity. The architecture of populist governing takes the form of stakeholder participation or forms of participatory governance that operate beyond the state and permit a form of self-management, self-organization, and controlled self-disciplining (see Dean, 1999; Swyngedouw, 2005; Lemke, 1999; Crouch, 2004), under the aegis of a nondisputed liberal capitalist order.

Third, populism customarily invokes the specter of annihilating apocalyptic futures if no direct and immediate action is taken. The classic racist invocation of Enoch Powell's notorious 1968 "Streams of Blood" speech to warn of the imminent dangers of unchecked immigration into the United Kingdom has, of course, become the emblematic populist statement, as are many of the apocalyptic headlines assembled in Table 1.1. If we refrain from acting (in a technocratic–managerial manner now), our world's future is in grave danger.

Fourth, populist tactics do not identify a privileged subject of change (like the proletariat for Marx, women for feminists, or the "creative class" for competitive capitalism) but instead invoke a common condition or predicament, the need for common humanity-wide action, mutual collaboration, and cooperation. There are no internal social tensions or internal generative conflicts. Instead, the enemy is always externalized and objectified. Populism's fundamental fantasy is of an intruder or, more usually, a group of intruders who have *corrupted* the system. CO_2 stands here as the classic example of a fetishized and externalized foe that requires dealing with if sustainable climate futures are to be attained. Problems therefore are not the result of the "system," of unevenly distributed power relations, of the networks of control and influence, of rampant injustices, or of a fatal flow inscribed in the system, but are blamed on an outsider. That is why the solution can be found in dealing with the "pathological" phenomenon, the resolution for which resides in the system itself. It is not the system that is the problem, but its pathological syndrome (for which the cure is internal). While CO_2 is externalized as the socioclimatic enemy, a potential cure in the guise of the Kyoto principles is generated from within the market functioning of the system itself. The "enemy" is, therefore, always vague, ambiguous, socially empty or vacuous, and homogenized (like "CO_2"); the "enemy" is a mere thing, not socially embodied, named, and counted.

Fifth, populist demands are always addressed to the elites. Populism as a project always addressed demands to the ruling elites; it is not about changing the elites, but calling the elites to undertake action. A nonpopulist politics is precisely about obliterating the elite, imagining the impossible, nicely formulated in the following joke: "An IRA man in a balaclava is at the gates of heaven when St. Peter comes to him and says, 'I'm afraid I can't let you in.' 'Who wants to get in?' the IRA man retorts. 'You've got 20 minutes to get the fuck out.' "

Sixth, no proper names are assigned to a postpolitical populist politics (Badiou, 2005b). Postpolitical populism is associated with a politics of not naming in the sense of giving a definite or proper name to the domain or field of action. Only vague concepts like climate change policy, biodiversity policy, or a vacuous sustainable policy replaces the proper names of politics. These proper names, according to Rancière (1995; see also Badiou, 2005a), are what constitute a genuine democracy, that is, a space where the unnamed, the uncounted, and, consequently, unsymbolized become named and counted. Consider, for example, how class struggle in the 19th and 20th centuries was exactly about naming the proletariat, its counting, symbolization, and consequent entry into the technomachinery of the state.

Seventh, populism becomes expressed in particular demands (get rid of immigrants, reduce CO_2) that remain particular and foreclose universalization as a positive socioenvironmental project. In other words, the environmental problem does not posit a positive and named socioenvironmental situation, an embodied vision, a desire that awaits its realization, a fiction to be realized. In that sense, populist tactics do not solve problems but merely move them around. Consider, for example, the current argument over how the nuclear option is again portrayed as a possible sustainable energy future and as an alternative to deal both with CO_2 emissions and peakoil. It hardly arouses the passions for what sort of better society might arise from this.

In sum, postpolitical postdemocracy rests, in its environmental guise, on the following foundations. First, the social and ecological problems caused by modernity/capitalism are external side effects; they are not an inherent and integral part of the relations of gobal neoliberal capitalism. Second, a strictly populist politics emerges here, one that elevates the interests of an imaginary "the People," "Nature," or "the environment" to the level of the universal rather than aspiring to universalize the claims of particular socionatures, environments, or social groups or classes. Third, these side effects are constituted as global, universal, and threatening: they are a total threat. Fourth, the "enemy" or the target of concern is thereby of course continuously externalized and disembodied. The "enemy" is always vague, ambiguous, unnamed and uncounted, and ultimately empty. Fifth, the target of concern can be managed through a consensual dialogical politics, and, consequently, demands become depoliticized.

PRODUCING NEW ENVIRONMENTS: A POLITICS OF SOCIONATURES

A true politics for Jacques Rancière (but also for others like Badiou, Žižek, or Mouffe) is a democratic political community, conceived as

> a community of interruptions, fractures, irregular and local, through with egalitarian logic comes and divides the police community from itself. It is a community of worlds in community that are intervals of subjectification: intervals constructed between identities, between spaces and places. Political being-together is a being-between: between identities, between worlds. . . . Between several names, several identities, several statuses. (Rancière, 1998: 137–138)

diversity key

Rancière's notion of the political is characterized in terms of division, conflict, and polemic (Valentine, 2005: 46). Therefore, "Democracy always works against the pacification of social disruption, against the management of consensus and 'stability.' . . . The concern of democracy is not with the formulation of agreement or the preservation of order but with the invention of new and hitherto unauthorised modes of disaggregation, disagreement and disorder" (Hallward, 2005: 34–35). The politics of sustainability and the environment, therefore, in their populist postpolitical guise are the antithesis of democracy, and contribute to a further hollowing out of what for Rancière and others constitutes the very horizon of democracy as a radically heterogeneous and conflicting phenomenon.

Therefore, as Badiou (2005b) argues, a new radical politics must revolve around the construction of great new fictions that create real possibilities for constructing different socioenvironmental futures. To the extent that the current postpolitical condition, which combines apocalyptic environmental visions with a hegemonic neoliberal view of social ordering, constitutes one particular fiction (one that in fact forecloses dissent, conflict, and the possibility of a different future), there is an urgent need for different stories and fictions that can be mobilized for realization. This requires foregrounding and naming different socioenvironmental futures, making the new and impossible enter the realm of politics and democracy, and recognizing conflict, difference, and struggle over the naming and trajectories of these futures. Socioenvironmental conflict, therefore, should not be subsumed under the homogenizing mantle of a populist environmentalist sustainability discourse, but rather should be legitimized as constitutive of a democratic order.

In the final paragraphs of this chapter I outline what constitutes for me the key drivers of conflict and where the possibilities for different fictions, and consequently new environmental futures, might tentatively reside.

The processes of sociophysical entanglement (what Marx called metabolic circulation) transform both social and physical environments and produce specific differentiated and unique social and physical milieus with new and distinct qualities (Swyngedouw, 2006). In other words, environments are combined sociophysical constructions that are actively (both by humans and nonhumans) and historically produced, both in terms of social content and physical-environmental qualities. Whether we consider the making of urban parks, natural reserves, or skyscrapers, they each contain and express fused sociophysical processes that embody particular metabolic and social relations (see Heynen, Kaika, & Swyngedouw, 2006). There is, in this sense, no single Nature, no One-All, but rather a great variety of distinct and often radically different (if not antagonistic) natures. There is nothing a priori unnatural or unsustainable, therefore, about produced environments like cities, genetically modified organisms, dammed rivers, or irrigated fields. The world is a Cyborg world, partly natural and partly social, partly technical and partly cultural, but with no clear boundaries, centers, or margins. The type and character of physical and environmental change and the resulting socioenvironmental flows, networks, and practices are not independent of the specific historical social, cultural, political, or economic conditions and the institutions that accompany them. All sociospatial processes are invariably also predicated upon the circulation and metabolism of physical, chemical, or biological components. Nonhumans, of course, play an active role in mobilizing socionatural circulatory and metabolic processes. It is these circulatory conduits that link often distant places and ecosystems together and permit relating local processes with wider sociometabolic flows, networks, configurations, and dynamics. These socioenvironmental metabolisms produce a series of both enabling and disabling socioenvironmental conditions.

Of course, such produced milieus often embody contradictory or conflicting tendencies. While environmental (both social and physical) qualities may be enhanced in some places and for some humans and nonhumans, they often lead to a deterioration of social, physical, and/or ecological conditions and qualities elsewhere. Processes of metabolic change are, therefore, never socially or ecologically neutral. This results in conditions under which particular trajectories of socioenvironmental change undermine the stability or coherence of some social groups, places or ecologies, while their sustainability elsewhere might be enhanced. Social power relations (whether material or discursive, economic, political, and/or cultural) through which metabolic circulatory processes take place are particularly important. It is these power geometries, the human and nonhuman actors, and the socionatural networks carrying them that ultimately decide who will have access to or control over, and who will be excluded from, access to or control over resources or other components of the environment and who or what will be positively or negatively enrolled in such metabolic imbroglios. These power geometries, in turn, shape the particular social and political configurations and the environments in which we live. Henri Lefebvre's (1974) "Right to the City" also invariably implies a "Right to Metabolism" (Swyngedouw, 2006).

Questions of socioenvironmental sustainability are fundamentally political questions revolving around attempts to tease out who (or what) gains from and who pays for, who benefits from and who suffers (and in what ways), from particular processes of metabolic circulatory change. Such politicization seeks answers to questions about what or who needs to be sustained and how this can be maintained or achieved. This includes naming socioenvironmental trajectories and enrolling them in a political process that is radically differentiated and oppositional. Clearly, Bush's notion of and desire for sustainability is not that of a Chinese peasant, a maquiladora woman worker, or a Greenpeace activist. It is important to unravel the nature of the social relationships that unfold between individuals and social groups and how these, in turn, are mediated by and structured through processes of socioecological change. In other words, environmental transformation is not independent from class, gender, ethnic, or other power struggles. Socioecological "sustainability" can only be achieved by means of a democratically (in the sense of a genuine political space) organized process of socioenvironmental (re)construction. The political program is to enhance the democratic content of socioenvironmental construction by means of identifying the strategies through which a more equitable distribution of social power and a democratically more genuine mode of the production of natures can be achieved.

A radical socioenvironmental political program, therefore, has to crystallize around imagining new ways to organize processes of socio-metabolic transformation. This requires first of all a radical repoliticization of the "economic," as it is exactly the latter that structures sociometabolic processes. But this is predicated upon traversing the fantasy that the "economic" is the determining instance of the political. Recapturing the political means foregrounding the political arena as the decisive material and symbolic space, as the space from which different socioenvironmental futures can be imagined, fought over, and constructed. This, of course, turns the question of sustainability radically to a question of democracy and the recuperation of the horizon of democracy as the terrain for the cultivation of conflict and the naming of different socioenvironmental futures.

NOTE

1. See *http://www.statistics.gov.uk* (accessed August 30, 2006).

REFERENCES

Agamben, G. (2005). *State of exception*. Chicago: University of Chicago Press.
Badiou, A. (2005a). *Metapolitics*. London: Verso.
Badiou, A. (2005b). Politics: A non-expressive dialectics. In *Is the politics of truth still thinkable?* A conference organized by Slavoj Žižek & Costas Douzinas, Birkbeck Institute for the Humanities, University of London.
Beck, U. (1994). The reinvention of politics: Towards a theory of reflexive modernization. In U. Beck, S. Lash, & A. Giddens (Eds.), *Reflexive modernization: Politics, tradition and aesthetics in the modern social order* (pp. 1–55). Cambridge, UK: Polity Press.
Beck, U. (1997). *The reinvention of politics: Rethinking modernity in the global social order*. Cambridge, UK: Polity Press.

Canovan, M. (1999). Trust the people!: Populism and the two faces of democracy. *Political Studies, 47*, 2–16.

Crouch, C. (2004). *Post-democracy.* Cambridge, UK: Polity Press.

Davis, M. (2006). Who is killing New Orleans? *The Nation, 282*, 14.

Dean, M. (1999). *Governmentality: Power and rule in modern society.* London: Sage.

Diken, B. L., & Laustsen, C. B. (2004). *7/11, 9/11, and post-politics.* Unpublished manuscript, Department of Sociology, Lancaster University, Lancaster, UK.

Giddens, A. (1991). *Modernity and self identity.* Cambridge, UK: Polity Press.

Giddens, A. (1994). *Beyond left and right.* Cambridge, UK: Polity Press.

Giddens, A. (1998). *The third way.* Cambridge, UK: Polity Press.

Hallward, P. (2005). Jacques Rancière and the subversion of mastery. *Paragraph, 28*(1), 26–45.

Heynen, N., Kaika, M., & Swyngedouw, E. (Eds.). (2005). *In the nature of cities: The politics of urban metabolism.* London: Routledge.

Laclau, E. (2005). *On populist reason.* London: Verso.

Latour, B. (2004). *Politics of nature: How to bring the sciences into democracy.* Cambridge, MA: Harvard University Press.

Latour, B. (2005). *Reassembling the social: An introduction to actor–network–theory.* Oxford, UK: Oxford University Press.

Lefebvre, H. (1974). *Le droit à la ville.* Paris: Ed. Seuil.

Lemke, T. (1999). The birth of bio-politics—Michel Foucault's lectures at the College de France on neo-liberal governmentality. *Economy and Society, 30*(2), 190–207.

Levene, M. (2005). Rescue!history: A manifesto for the humanities in the age of climate change—an appeal for collaborators. Retrieved September 12, 2006, from *http://www.crisis-forum.org.uk/rescue_history.htm.*

Lomborg, B. (1998). *The skeptical environmentalist: Measuring the real state of the world.* Cambridge, UK: Cambridge University Press.

Marvin, S., & Medd, W. (2006). Metabolisms of obe*city*—Flows of fat through bodies, cities and sewers. In N. Heynen, M. Kaika, & E. Swyngedouw (Eds.), *In the nature of cities—The politics of urban metabolism* (pp. 161–179). London: Routledge.

Mouffe, C. (2005). *On the political.* London: Routledge.

Rancière, J. (1995). *La mésentente: Politique et philosophie.* Paris : Editions Galilée.

Rancière, J. (1998). *Disagreement.* Minneapolis: University of Minnesota Press.

Rancière, J. (2003a). Comment and responses. *Theory and Event, 6*(4).

Rancière, J. (2003b). Ten theses of politics. *Theory and Event, 5*(3).

Rancière, J. (2005a). *Chroniques des temps consensuels.* Paris: Seuil.

Rancière, J. (2005b). *La haine de la démocratie.* Paris: La Fabrique.

Stengers, I. (2003). *Cosmopolitiques.* Paris: La Découverte.

Swyngedouw, E. (2005). Governance innovation and the citizen: The Janus face of governance-beyond-the-state. *Urban Studies, 42*(11), 1991–2006.

Swyngedouw, E. (2006). Circulations and metabolisms: (Hybrid) natures and (cyborg) cities. *Science as Culture, 15*, 105–121.

Thomson, A. J. P. (2003). *Re-placing the opposition: Rancière and Derrida. Fidelity to the disagreement.* Unpublished manuscript, Goldsmith's College, University of London.

Urbinati, N. (2003). Can cosmopolitan democracy be democratic? In D. Archibugi (Ed.), *Debating Cosmopolitics* (pp. 67–85). London: Verso.

Valentine, J. (2005). Rancière and contemporary political problems. *Paragraph, 28*(1), 46–60.

Žižek, S. (1998). For a leftist appropriation of the European legacy. *Journal of Political Ideologies, 3*(1), 63–78.
Žižek, S. (1999a). Carl Schmitt in the age of post-politics. In C. Mouffe (Ed.), *The Challenge of Carl Schmitt* (pp. 18–37). London: Verso.
Žižek, S. (1999b). *The ticklish subject: The absent centre of political ontology.* London: Verso.
Žižek, S. (2002a). *Looking awry: An introduction to Jacques Lacan through popular culture.* Cambridge, MA: MIT Press. (Original work published 1992)
Žižek, S. (2002b). *Revolution at the gates: Žižek on Lenin—the 1917 writings.* London: Verso.
Žižek, S. (2005a). Against human rights. *New Left Review, 34,* 115–131.
Žižek, S. (2005b). Against the populist temptation. In *Is the politics of truth still thinkable?* A conference organized by S. Žižek & C. Douzinas, Birkbeck College, University of London, The Birkbeck Institute for the Humanities.
Žižek, S. (2006). *The parallax view.* Cambridge, MA: MIT Press.

CHAPTER 2

Sustaining Modernity, Modernizing Nature

The Environmental Crisis and the Survival of Capitalism

ROGER KEIL

This chapter examines the relationships between the theory and practice of two related concepts: sustainability and ecological modernization. It will be argued that these are both complicit in a process through which capitalist development has been increasingly linked to ecological concerns. Some of this development has roots that reach as far back as the 1960s, but it can be more directly traced back to the advance of both theories and practices of ecological modernization in the 1990s. "Sustainable development" was the slogan of the late 1980s, and "ecological modernization" has since become an omnipresent and sometimes dominant neoliberal approach for greening capitalism. As 21st-century economies search for ways to reinvent themselves in the face of mounting evidence of global environmental change, environmental theory has, in large part, fallen into line with capitalist hegemony. Some may even argue that green global capitalism may be the precondition for the survival of the system overall. Tradable permit schemes, green taxes and fees, and other market-oriented ways of dealing with the inevitable fallout of industrial society in both its material/ecological and social/cultural dimensions are the favored modes of environmental protection in a post-Kyoto world (Desfor & Keil, 2004). Yet, as the joint processes of globalization and neoliberalization continue to take their toll on communities and ecologies around the globe, it becomes ever more doubtful whether ecological modernization and sustainability schemes will hold the test of practice. Cynicism and blanket dismissal of all notions and practices of ecomodernization and sustainability are clearly out of place. There is a broad spectrum of new ideas and activities that associate with both and that have created openings for an improvement of the societal relationships with nature. Things can be worse than the win–win situations proposed by ecological modernization, and in many instances they still are. While it is argued in this chapter that ecological modernization owes much to neoliberal ideas, there are other, more dangerous, neoliberal practices that continue down the older route of shameless exploitation of natural resources and human communities without any recognition of any limits to either. Still, I will treat ecological modernization and sustainability as largely encouraged and emboldened by neoliberalization. In fact, their hegemonic character in the regulation of human–natural relationships has been based on their compatibility with the neoliberal project, particularly in western Europe and parts of North America but also in parts of the developing world (Bond, 2000, 2002; York & Rosa, 2003). In this sense, ecological modernization justly draws upon itself a rigorous analysis by critical environmental theorists and practitioners, who are both wary of its promises and dismayed by the environmental movement's inability to counter the new hegemony's obvious appeal. The overwhelming noise produced by ecomodernization has called forth a new generation of concerned environmentalist thinkers who have noted the fallacy of our belief that we can develop our way out of capitalism. At one end of the spectrum of this new thinking is the political economic Marxist critique of capitalism, which has pointed to the necessarily destructive and entropic tendencies in global capitalism (Altvater, 1993; Harvey, 1996; Kovel, 2002; O'Connor, 1998). At the other end is a more culturally informed critique of the environmental movement itself. This critique, exemplified in the brilliant work of Frederick Buell (2003), exposes the increasing bluntness of the traditional "apocalyptic" environmental critique in the face of ongoing ecological destruction. As Buell argues perceptively, while the world collapses around us, we have gotten used to the ambient noise that accompanies that collapse. In fact, we have not only learned to tolerate the persistent messages on the end of the world as we know it, we have come to revel in that prospect.

The chapter proceeds with theoretical reconsideration of the concept of the "societal relationships with nature" (for an overview, see Görg, 2003) to argue that the current (or persisting) crisis of these relationships needs to be understood as the basis of the sustainability problematique. It will be proposed that all such relationships are both material and symbolic and that their sustainability relies on these two dimensions to a largely equal degree. It will finally be argued further that these relationships now are particularly relevant in the urban context where much of today's socioecological problems are produced but where also many of the solutions to these problems will have to be found. The chapter ends with a proposal for an urban political ecology, which takes as its starting point the notion of the crisis of the societal relationships with nature but points to sustainability as an elusive goal as long as fundamental processes of uneven development are not brought under control. This unevenness now has to be seen in the global biopolitical context of empire, which perpetuates itself as an urban-based yet globally calibrated project that is destructive during times of peace and entirely out of control during times of war (Hardt & Negri, 2000, 2004).

LIMITS TO GROWTH: THE CRISIS UNFOLDS

Although the alarm bells had tolled as early as 1972, when the Club of Rome published its striking *The Limits to Growth*, sustainability in the way we are using it today was born from the fundamental crisis in world economies between 1974 and 1982 and the impasse in global development during the 1980s. Specifically, as readers of this book will know well, the term "sustainable development" was coined by the so-called Brundtland report of the World Commission on Environment and Development, which famously defined it as "development that meets the needs of the present without jeopardizing the ability of future generations to meet their own needs." Sustainability was introduced as a concept of societal modernization on a global scale. It was operationalized during the 1980s and 1990s as a master concept of societal change in an increasingly neoliberal world.

Both the Club of Rome and the Brundtland reports were potentially revolutionary, as they proposed an immanent critique of capitalism. Once the genie was out of the bottle—once growth itself was identified as a problem—capitalism could not easily be put together again as a working model despite the fact that after 1990 in particular "there appear[ed] no attractive alternatives to capitalism left" (Becker & Jahn, 1998: 75). Still the question remained: "Will the social wealth produced under capitalist conditions perhaps bring final destruction to the natural basis for the life of humanity—or is there a realistic possibility of civilizing capital both socially and ecologically? The answer to this question will decide whether sustainable development is merely an ideological slogan or a signpost pointing the way to a new concrete utopia" (Becker & Jahn, 1998: 75).

As Becker and Jahn (1998) showed, the Club of Rome report had identified "economic growth as a 'world *problematique*' spanning societies from the First to the Third World, both capitalist and state socialist countries." Focusing on the fatal combination of population growth and economic growth in particular, the system-transcending warning that emanated from the scientists of the Club of Rome was that humanity might be en route to "destroying the still open path to a society of the free and equal, and driving us to disaster if not halted." Yet, the "growth-limiting world *resolutique* adapted to the world *problematique*" proved elusive as capitalist growth, the growth of global populations, and the use of resources spun out of control (Becker & Jahn, 1998: 69).

But was there a grand anticapitalist ecological awakening? No. What happened instead, of course, was the turbo-response of neoliberalism. While both the Club of Rome and the Brundtland reports were answers to a specific set of issues highlighted by half a century of Fordist industrial megagrowth and failed global development (trying to replicate the American model worldwide with disastrous consequences), the proposed solutions for reform of the world capitalist system were born in a distinctly social democratic era of Keynesian interventionism and growing social equity in Western societies. The Club of Rome report was mostly effective in confronting the Fordist mass-consuming societies of the West with their own irrationalities and fatal trajectories. It laid the groundwork for the kind of sensitivities—in concert with other, more radical, ecological, neo-Marxist, and feminist critiques—that created strong Green party movements in many European countries as well as (to a lesser degree) in Canada and the United States.

The Club's proposals were advanced, ultimately, by a group of 21 international world luminaries under the leadership of the former social democratic Norwegian prime minister Gro Harlem Brundtland. Whereas the Club of Rome had targeted the sensitivities of an out-of-control consumer society at the height of Fordism, Brundtland's World Commission on Environment and Development addressed the West's growing fears that development as it was known after 1945 was not delivering the goods: economic growth in an era of accelerated neocolonialism and imperialism was stalled as countries of the global South were mired in a debt crisis of unknown proportions, ecological problems abounded as populations still exploded (filling up the shantytowns of the large cities), and violence was endemic inside and between developing nations as independence as well as the incipient dissolution of the Cold War blocs had set free the centrifugal dynamics of militarism, civil wars, and permanent revolutions.

The Brundtland Commission operated on a truly planetary level and pointed to the destructive dialectics of individualist self-interested activity in a shared world as a major problem. Under the slogan "The Earth is one but the world is not," Brundtland discussed the need for concerted worldwide action to stave off the detrimental effects of competitive striving for "survival and prosperity with little regard for its impact on others." It also identified poverty as the central pivot around which the question of further development had to be articulated. The apocalyptic tone of the Club of Rome's report was perpetuated in statements that stressed the real possibility of irreversible damage to the world's environments (and subsequently human progress) and the "deepening interconnections" of human and natural environments worldwide. The report isolated two key concepts:

- The concept of "needs," in particular the essential needs of the world's poor, to which overriding priority should be given; and
- The idea of limitations imposed by the state of technology and social organization on the environment's ability to meet present and future needs. (World Commission on Environment and Development, 1987)

In language that has now penetrated world discourse, from the offices of the United Nations to primary school science classes, from business ethics seminars to local governments, and from the recycling centers of the North to the policies of nongovernmental development organizations in the South, the Brundtland report created the blueprint for the "world resolutique" that had been elusive to the authors of the Club of Rome report. Especially the politicization of sustainability after Rio 1992 in a civil society and local community-based political process, which shifted responsibility for sustainability from the global capitalist corporations to individuals, their communities, their bodies, and personal metabolism, led to an ongoing global environmental pragmatism in myriad forms and at all scales of human activity around the globe.[1]

The Brundtland and Club of Rome proposals were subsequently recast from a critical set of potentially anticapitalist warnings to a recipe for the survival of capitalism through a concrete set of measures involving ecological modernization. Instead of throwing a wrench into the capitalist machine, sustainability subsequently gets redefined as one of the possible routes for a neoliberal renewal of the capitalist accumulation process. This was particularly successful, as it proved possible in the years after the Rio conference (and despite the total neglect of that process in the world's only remaining superpower, the United States) to turn "sustainable development," and even more its less oxymoronic sister, "sustainability," into the new governmentality of a neoliberalized global capitalism. It was precisely the internalized individual and community-focused responsibility of the neoliberal era that undermined both the potentially fundamental critique of capitalist development and the search for social solutions to the socioenvironmental crises that kept unfolding worldwide.

Following a decade of postmodern uncertainties and fractious partialities, a new master discourse seems to be emerging based on the concept of "ecological modernization." Debates about this new discourse, largely dominated by Europeans, have become a beacon of hope to many during a period of rapid change in a world system driven by globalization and neoliberalization. Ecomodernizers speak of win–win situations, manageable futures, and prosperous development *with* rather than *against* "nature." The making of environmental policies plays an important part as states restructure themselves in scales and functions. Within this context, many policy analysts assume that sustainable futures can be attained under conditions of a continuously growing capitalist economy by making use of negotiated problem-specific settlements among different and divergent policy actors. The relative success of ecological modernization in the 1990s in European, and to a lesser degree in North American, societies has led to the common perception that capitalism has found a reliable mode to green itself. Ecomodernized capitalism, it is claimed, can be counted on to take good care of our planet. In this sense, ecological modernization is part of an overall approach toward sustainability and, more particularly, about nature under capitalism. In contrast to radical critics of capitalism's destructive forces, proponents of ecological modernization tend to point to the real (and some would even argue "proven") capacities of capitalism to change and become more environmental, to produce and consume in environmentally sound ways, and, indeed, there is no serious alternative to capitalism as we know it.

This kind of reasoning, Frederick Buell would argue, is on a dangerous slope of confusing the strategic and necessary goal of environmentalism—the protection of first nature—with its more expendable companion, the discourse about the relationships with that nature. In fact, as Buell argues, the pervasive problematizing of nature as constructed makes us numb as we begin to take the destruction of first nature for granted while we are fiddling with the symbolic and cultural discourses that explain but ultimately obscure the material, physical, and ecological destruction around us. *Talk* about sustainability, Buell seems to argue, seems to keep us from the dramatic actions needed to actually sustain the natural processes that keep us alive in the long term. Buell takes us back to the pre-Club of Rome call to arms by Rachel Carson, who announced to an unsuspecting post-World War II generation that we live in a "sea of carcinogens." This had been shocking news to those who had just settled comfortably into the guiltless joys of consumerism, bolstered by exuberant belief in the capacity of humanity to conquer nature, to split atoms, feed the hungry, send rockets to the moon, and reverse the flow of water. That this notion of living in crisis is *not* shocking news anymore is both the beginning and end of Buell's detailed analysis of the politics, discourses, and literatures of environmentalism. If we have learned to live in apocalypse, one may argue by extension that we certainly have also learned to look at it. In this context, sustainability becomes an elusive objective.

Working through theoretical developments such as ecological modernization and the risk society as well as political tendencies such as attempts to green and globalize American politics under Clinton and Gore, Buell seeks to demonstrate that we are entwined in a double process of hypermodernization and postmodernization and that we have subscribed to a popular and scientific culture of hyperexuberance, where the likes of antienvironmentalist Julian Simon, the proponents of chaos theory, and computerization with its WIRED culture have us transfixed in a worldview that accepts crisis as normal. Buell spares some of his most outspoken criticism for pragmatic ecological modernizers and sustainability advocates, about whom he concludes: "And even the most positive assessment would have to conclude that there is still, alas, a wide gap between actual achievement and the fact that 'ecological modernization has produced a real change in *thinking* about nature and society in the *conceptualization* of environmental problems in the circles of government and industry' " (Buell, 2003: 51, citing Hays, 1987: 250).

Richard York and Eugene Rosa would agree with Buell's assessment. In their examination of the claims of ecomodernization, which they identify as a "prominent neoliberal theory," they come to this conclusion: ecological modernization theory "has failed thus far to provide a convincing case that late modernization is compatible with, let alone essential to, the development of ecological sustainability" (York & Rosa, 2003: 281). Others have argued that ecological modernization theory and practice have been hegemonic in the face of challenger environmentalisms such as urban and social ecology as well as environmental justice (Desfor & Keil, 2004). Buell's cultural critique of ecomodernization and sustainable development is echoed in radical social critiques, particularly among the eco-Marxist scholars who have grouped around the journal *Capitalism, Nature, Socialism*. While there is much diversity and debate in this journal, its editor, Joel Kovel, speaks for many when he identifies capitalism as "the enemy of nature" (2002). Speaking from the point of view of a critique of capitalist development, Kovel explains:

> From this standpoint, the ecological crisis may be said to be human production gone bad. Put more formally, the current stage of history can be characterized by *structural forces that systematically degrade and finally exceed the buffering capacity of nature with respect to human production, thereby setting into motion an unpredictable yet interacting and expanding set of ecosystem breakdowns.* The ecological crisis is what is meant by this phase. In it we observe the desynchronization of lifecycles and the disjointing of species and individuals, resulting in the fragmentation of ecosystems human as well as non-human. For humanity is not just the perpetrator of the crisis: it is its victim as well. And among the signs of our victimization is the incapacity to contend with the crisis or even to become conscious of it. (Kovel, 2002: 21, emphasis in the original)

As to the ability for capitalism to provide the goods (wealth for all) and to not destroy the environment in the process, Kovel is clear: "Poverty, eternal strife, insecurity, ecodestruction and, finally, nihilism are also produced. . . . The ecological crisis is the name for the global ecodestabilization accompanying global accumulation" (Kovel, 2002: 82).

Radical political ecologists have mostly agreed with this assessment. While there has been some debate among progressive ecologists on how much of the ecological crisis has progressed beyond the causality of capital accumulation and about the relative value of terms such as "scarcity" and "resource depletion" (fearing Malthusian slips in the process), there has been growing concern in this literature for the future of ecosystems and the survivability of the human species in a world of accelerated entropy and poisoned metabolisms (Altvater, 1993; Harvey, 1996; Heynen, Kaika, & Swyngedouw, 2006; Lipietz, 1992). In recent years, more specific assessments of the capitalism–nature relationships have dissected the specific relationships of neoliberal capitalism to an increasingly privatized and commercialized nonhuman nature (see special issues of *Capitalism, Nature, Socialism* on neoliberalism and nature in the Spring and Summer issues of 2005). But as the critiques zoom in on capitalism as the enemy of nature, sustainability as an accumulation strategy is also being discounted. The introduction to a recent collection of critical papers on sustainability states clearly: "Inadequate responses to ecological exhaustion abound. Corporate elites and most governments promote a form of sustainability that is more about sustaining capitalism, growth, and profits than sustaining living environments" (Johnston, Gismondi, & Goodman, 2006b: 13). But from the ensuing devastation, the same authors also draw the possibility of a new politics in reaction to the privatization and neoliberalization of nature: "Such incursions open new political spaces and opportunities for social and political movements to contest corporate applications and meanings of sustainability" (Johnston et al., 2006b: 14). The authors offer a critical view on the commonly held "human/nature separation obscuring the unavoidable connection between ecology and human existence" (Johnston et al., 2006b: 15–16). They provide four "conceptual maps"—the ecosocial crisis, maldevelopment and ecoimperialism, commodification and biosphere degradation, and ecopolitics and geography—and trace both the systemic contradictions of capitalism's continued ravaging of nature and the potential progressive political opportunities that arise from these contradictions. Before we proceed down this route, however, we will have to look more deeply into the abyss that seems to have opened up under human societies in the natural world.

BEYOND SUSTAINABILITY?: THE MILLENNIUM ASSESSMENT

As far as global environmental reports go, the ongoing Millennium Ecosystem Assessment under the stewardship of the United Nations has certainly ratcheted up the ante. This exercise, which involves more than 1,300 scientists worldwide, is the most extensive attempt by a global agency yet to take the ecological pulse of the planet, and the results so far point to almost insurmountable obstacles to sustainable development unless drastic changes are implemented in human activities in the very near future. As global scientific consensus has grown on central issues such as the shrinkage of biodiversity and climate change, there is now less discussion about *if* than about *when* the ecological crisis will overwhelm world society. The key messages sent by the authors of the report add up to a stern warning to human society that their impact on nonhuman nature might be at a potentially irreversible point from where any type of sustainability action might be hard to imagine. The report acknowledges the increases in wealth and well-being and improvements in "the lives of billions, but at the same time they weakened nature's ability to deliver other key services such as purification of air and water, protection from disasters, and the provision of medicines" (Millennium Ecosystem Assessment, 2005: 3). Under the heading "the bottom line" the report states unequivocally:

> At the heart of this assessment is a stark warning. Human activity is putting such strain on the natural functions of the Earth that the ability of the planet's ecosystems to sustain future generations can no longer be taken for granted. . . . Above all, protection of these [natural] assets can no longer be seen as an optional extra, to be considered once more pressing concerns such as wealth creation or national security have been dealt with. This assessment shows that healthy ecosystems are central to the aspirations of humankind. (Millennium Ecosystem Assessment, 2005: 5)

While based on the most extensive *scientific* global base to define the sustainability problematique, the Millennium Ecosystem Assessment makes direct and unambiguous recommendations for the rearrangement of the societal relationships with those natures that are under pressure. The report notes that "coordinated efforts across all sections of government, businesses, and international institutions" would be required to stave off the dire ecological future it predicts if no action is taken. No doubt, this unprecedented and important document and the process that generated it are authoritative in a global society that seeks to come to terms with its political, ideological, socioeconomic, and cultural fragmentations on more fronts than the ecological balance of the planet. However, it suffers from two interrelated flaws:

1. It perpetuates the "modern" scientific constitution of separating "science" from "politics" which has been identified by Latour (2004) and which may itself be part of the problem. In a discussion of this constitution, which cannot be taken up here in full, Latour (2004) has argued that

> the terms "nature" and "society" do not designate domains of reality; instead, they refer to a quite specific form of public organization. . . . By dividing public life into two incommensurable houses, the old Constitution led only to paralysis, since it achieved only premature unity for nature and endless dispersions for cultures. The old Constitution thus finally resulted in the formation of *two equally illicit assemblies*: the first, brought together under the auspices of Science, was illegal, because it defined the common world without recourse to due process; the second was illegitimate by birth, since it lacked the reality of the things that had been given over to the other house and had to settle for "power relations," for a multiplicity of irreconcilable viewpoints, for Machiavellian cleverness alone. (53–54)

Latour's critique cuts deep and is equally trenchant for both the classical positivist scientific method that ruled the hard facts of science out of reach of the political and for those among the social constructivists, who believe most strongly in the rule of a separated social over the natural. The latter include of course the neoclassical economists as much as the conventional Marxists, who believed in the ability of humankind to take control over natural processes through labor. Latour's intervention of "reassembling the social" points beyond both conventional tendencies: positivist science, both social and natural, as well as conventional materialist dialectic, which prioritizes human demands over the ecological actor-networks that characterize the world in which we live (Latour, 2005; for a Marxist application of Latourian thought, see Swyngedouw, 2004). In effect, the Millennium report presents the practices and results of rational scientific endeavor as unproblematic and takes its outcomes and recommendations as positive facts while paying scant attention to the implication of scientific work in the production of the crisis itself. There is no need to lecture the world's leading scientists about the systemic imbrication of humans with nature. But it needs to be pointed out that the Millennium report does not break the "modern" mold of active autonomous humans upon a helpless and passive nature. Clearly, the results of the report are alarming and need the broadest possible publicity in human societies everywhere. Yet, the report also does not go beyond its predecessors, *The Limits to Growth* and *Our Common Future,* in its unflinching anthropocentrism and scientific positivism.

2. The report is rather oblivious to the real global rifts among and inside human societies and collectives, which stand in the way of concerted action. Most active human decision makers at the level of global leadership now publicly acknowledge the necessity of concerted action to enhance sustainability and to defeat ecological doomsday scenarios. Occasional political grandstanding about nuclear energy and weapons, water scarcity, and climate change aside, most people with any intellectual capacity don't need convincing that action is overdue. The problem has been that the world "resolutique" has been entirely off political agendas filled with bilateral, international, and regional conflicts in a thoroughly rescaled world of conflicted geographies (Sparke, 2005). Today's world is perhaps not quite as simplistically dualized as in Benjamin Barber's "Jihad against McWorld" (Barber, 1995). Yet, in an environment of empire on one hand and massive de- and reterritorializations on the other, the "world problematique" identified by the Millennium Assessment falls into a giant "institutional void" (Hajer, 2003) of a jurisdictional battle zone and democratic vacuum. A possible "global" solution as sought by the Millennium Assessment is elusive in the face of the real-world multiscale fragmentations of empire/multitude as experienced in real time as we continue down the path of ecological destruction outlined in the assessment. Clearly, there has been some movement in the positions of American presidents from George Bush I's infamous statement during the Rio 1992 summit that the American way of life was not up for discussion to the admission by his son George Bush II in 2006 that Americans had to shed their addiction to oil. But in fact the conditions under which action could be taken to follow through on such insights have worsened in the long decade that separates the two Bush presidencies as world politics has become more contentious along a number of lines in the period following the Seattle World Trade Organization meetings of 1999, the attacks of Al Quaeda on New York and Washington in September 2001 and the invasion of Iraq by the U.S.-led "coalition of the willing" in 2003. The weight of the Millennium Ecosystem Assessment report cannot be denied by any politics of sustainability. But as long as the report remains stuck in a polity established by the old purified political Constitution in Latourian terms and as long as it is oblivious to the real power relations in the age of empire, it will amount to little.

FROM THE CRISIS OF THE SOCIETAL RELATIONSHIPS WITH NATURE TO A POLITICAL ECOLOGY OF SUSTAINABILITY

There are no good arguments at this point in human history to dispel the sense of ecological catastrophe that surrounds us. Sustainability has taken a backseat to capitalist development. At the basis of the sustainability world problematique has been the growing rift between the global awareness of catastrophic failure—symbolized by the scientific consensus on global warming and the disturbing signs of climate change (hurricanes, droughts, melting polar ice caps, etc.) worldwide—and the cocky persistence of the neoliberal growth model under American leadership that theirs is the way of the future despite George W. Bush's recent declarations about America's oil dependency. This gap points ever more to the necessity of developing a different notion of crisis than is usually common in environmental circles. And perhaps it is through this redefinition that we find an intellectual and political pathway to prevent what seems to be the inevitable: the extinction of human society from the planet.

We return briefly to Becker and Jahn's critique of the growth/development model and have a closer look at the theoretical framework they employ in the process of identifying a "resolutique" that might actually deal with the problem identified by the Club of Rome, the Brundtland Commission or even the Millennium Ecosystem Assessment. Becker and Jahn's concept of "socio-ecological transformation" holds as a powerful starting point for the analysis needed to move the resolution of our growing ecological malaise forward. They note the imbrication of material with cultural aspects of our lives and highlight the fact that

> different societal relationships with nature must not only be materially regulated, but always culturally symbolized as well. The cultural regulation of sexuality and reproduction takes a privileged position, which, beside labour and production, constitute a second pole where the regulation of all other natural relationships condense symbolically in differences of sex and gender. Labour and production, eating and drinking, locomotion and reproduction are basic relationships on which others (such as clothing, shelter and protection against danger) depend. The various material forms of regulation have corresponding elements in a symbolic order—mediated via language, rite, myth, religion, art and science. And they determine the forms of individual participation in social life. (Becker & Jahn, 1998: 81)

Becker and Jahn's conclusion from this is important: "What is at the present time being discussed as the ecological crisis is in essence a crisis in the societal relationships with nature and problem solution means intervention in their dynamics—with often unpredictable and unwanted, dangerous side-effects" (Becker & Jahn, 1998: 81).

We can take from this insightful shift in perspective—from the "ecological crisis" to the "crisis in the societal relationships with nature"— the lesson that we need to look simultaneously at the symbolic and the material aspects of the sustainability problematique. The problem with Becker and Jahn's proposal as well as with similar ideas about the "eco-social crisis" (Johnston et al., 2006b) is the continued maintenance of the dichotomy of the natural and social as if they were two poles of a rather fixed universe. We need to first enlist Latour's insights, briefly explained above, about the modern constitution that underlies our current political life: the separation of the two "houses" of science and politics. The notion of "societal relationships with nature" seems, in fact, to perpetuate this separation even as it criticizes the dualistic view of nonsocial nature and nonnatural society. Instead of the old politics of nature and society that has also informed the politics of sustainability, Latour (2004) introduces a new notion of the political

> Political ecology is not going to be simpler, nicer, more rustic, more bucolic, than the old bicameral politics. It will be both simpler and more complicated: simpler because it will no longer live under the constant threat of a double short-circuit, by Science and by force, but also much more complicated, for the same reason—for want of short-circuits, it is going to have to start all over and compose the common world bit by bit. In other words, it will have to *engage in politics*, an activity to which we had finally gotten rather unaccustomed, given the extent to which confidence in Science had allowed us to postpone the day of reckoning in the belief that the common world had already been constituted, for the most part, under the auspices of nature. (pp. 82–83)

The contentious and confusing part of Latour's thinking is its extension of the legitimate range of agency to all manner of animate and nonanimate postcitizen actors, which he calls *propositions*.

> I am going to say that a river, a troop of elephants, a climate, El Niño, a mayor, a town, a park, have to be taken as propositions to the collective. . . . I do not seek to claim that the pluriverse is composed of propositions, but simply that in order to begin its civic work of collection, the

> Republic is going to consider only propositions instead of and in the place of the earlier subjects and objects. (Latour, 2004: 83)

We cannot propose Latour's new constitution as a *prêt-à-porter* recipe for the world resolutique sustainability must be looking for. But, together with Donna Haraway's notion of the cyborgian human existence, Latour helps us to assess the challenge of politics in a after-"modern" world where science and politics, reflection and action, cannot remain separated.

For the remainder of this chapter, I suggest that one arena where these relationships are negotiated in real terms is the urban, and I will make the case for a specifically urban political ecology that is mindful of Becker and Jahn's observations but which also goes beyond them.

GETTING PRACTICAL:
POLITICIZING THE SUSTAINABLE CITY

One of the tangible processes through which the destructive, unsustainable process of capital accumulation has expressed itself in material and discursive terms is urbanization. This process is contradictory in that it is both exhaustive of natural and human resources in unprecedented ways and potentially the one way in which we can sustain a human population that is projected to reach 10 billion by 2050. The first aspect has been decried not only by defenders of rural ways of life and conservative antiurbanists, who see city life as destructive of tradition and community and productive of physical decay, moral decrepitude, and ecological destruction, but has also been subject to progressive critiques, which have noted the increasingly catastrophic tendencies in mostly squatter-based human settlement around the world and the reproduction of spatial inequity on a world scale (Davis, 2005; Neuwirth, 2005). It is clear, though, that urban and regional environments have become an important scale at which a "sustainability fix" (While, Jonas, & Gibbs, 2004) has been sought through concerted municipal and regional policy efforts (Keil & Boudreau, 2006), and where a form of "real existing sustainability" has been achieved (Krueger & Agyeman, in press), although it is noteworthy that it important to retain a multiscale perspective on the sustainability problematique (Grenfell, 2006).

In viewing the urban dimension of sustainability, we must find the strongest words of criticism for the very common notion that we can plan sustainability through smart social engineering and urban design. In fact, urban planning, which has prided itself on iteratively changing our urban environments for the better, must be exposed to a ruthless critique. As a prominent example for the rather un-self-critical practice of sustainability planning stands a recent statement by Sir Peter Hall, arguably the world's most well-known living planner. Under the title "The Sustainable City: A Mythical Beast," Hall conjugates the history of modern city planning, arguing in essence that, using the register of planning developed during the 20th century, we will be able to achieve urban sustainability: "We have done sustainable cities. We have done them 100 years ago. We've done them 50 years ago. We've done them 30 years, and we have been doing them in the last decade. It can be done. It needs some money, including government money—for infrastructure, in particular. It needs power, in some cases, to carry them through. . . . It needs, above all, imagination and determination" (Hall, 2005: 17). These are good words and good intentions, but they are in grave disrespect of the true conditions under which cities and sustainability intersect. They are representative of an industry of sustainability planning that has flooded design practice and teaching with a new set of dogmas that are summed up in a growing number of manuals and manifestos for "cities for a small country" or "small planet" (Rogers & Gumuchdjian, 1997; Rogers & Power, 2000).

We can also set aside well-meaning but ultimately insufficiently far-reaching notions of the "ecological footprint" (Wackernagel & Rees, 1996) and urban sustainability as set forth in so many "how-to" books that propose many small steps toward greening communities but lose sight of the big picture of capitalist urbanization dynamics. We can also leave behind urban design proposals that are now conveniently summarized as "new urbanism" and "transit-centered development": they cannot reach deeply enough to fundamentally redirect the destructive dynamics of today's urbanism (represented in the architecture of Andres Duany and Elizabeth Plater-Zyberg and Peter Calthorpe). Moreover, an attempt to redirect the city's material streams from a linear to a circular metabolism, as suggested inter alia by Herbert Girardet (1992), is not a promising strategy to achieve sustainability so long as it leaves the social processes that symbolize and sustain these processes untouched. Lastly, we need to continue our critical—even if sympathetic—accompaniment of Local Agenda 21 processes both in the global South and North (Low, Gleeson, Elander, & Lidskog, 2000).

Instead, I would like to propose here a radical urban political ecology (UPE). Radical UPE has developed around three major themes. All of them are keenly constructivist in their epistemology and subscribe to an ontology that regards the capitalist accumulation process as the basis of the crisis of the societal relationships with nature. When they articulate themselves with sustainability agendas, they do so in the conviction that such sustainability can only be achieved at the expense of capitalism as we know it. The first theme is that of urban metabolism, the second is the notion of cyborg urbanization, and the last one is environmental justice.

The notion of urban metabolism has long been bandied about by urban ecological thinkers, but it has recently received more attention as critical Marxists have re-infused the term with meaning. Swyngedouw and Heynen (2003: 906–907) have noted the "interwoven knots of *social process, material metabolism* and *spatial form* that go into the formation of contemporary urban socionatural landscapes. . . . It is on the terrain of the urban that [the] accelerating metabolic transformation of nature becomes most visible, both in its physical form and its socio-ecological consequences." This use of the concept of metabolism is more complex and much different in its intellectual and political context than its traditional use in the urban field. Following Abel Wolman's seminal article "The Metabolism of Cities" (1965), a systems theoretical awareness of the material streams that keep the city (and its residents) alive had entered the imaginary of at least the environmentally interested (if often only the technologically minded) urban studies scholars. Still, given the importance of the topic for the everyday sustainability of urban regions, there has been comparatively little uptake in rigorous empirical studies of the metabolism of cities. The major (and lone) exception has been the studies of Hong Kong's metabolism (Newcombe, Kalma, & Aston, 1978; Warren-Rhodes & Koenig, 2001). On the conceptual front, a schematic and often polemical use of the concept has found much application—in classrooms and in popular debate among environmentalists—as a consequence of Herbert Girardet's widely popular use of "linear" and "circular" metabolism in his *Gaia Atlas of Cities* (Girardet, 1992) and in a more recent primer on the subject called *Creating Sustainable Cities* (Girardet, 1999). This work also ties in with the mostly didactic literature on the "ecological footprint" (Wackernagel & Rees, 1996), which stops short of applying critical political analysis to the state of the world's urban environments, as it concentrates on technological and behavioral solutions and treats metabolism as mostly a form of biophysical exchange. Despite its limitations, this work can provide an important grounding for discussions on urban political ecology. A recent study on the metabolism of Toronto, for example (Sahely, Dudding, & Kennedy, 2003), is of prime importance in understanding the precarious political ecologies this exploding metropolis relies upon. Sahely, Dudding, and Kennedy define urban metabolism as "a means of quantifying the overall fluxes of energy, water, material, and wastes in and out of an urban region. Somewhat analogous to human metabolism, cities can be analyzed in terms of their metabolic flow rates that arise from the uptake, transformation, and storage of materials and energy and the discharge of waste products" (Sahely et al., 2003: 469; Warren-Rhodes & Koenig, 2001). From a UPE point of view, we might add the following caveats to this project:

1. Beyond reference to policy changes (e.g., introduction of recycling), there is little attention paid in these studies to the political changes in the study area.
2. While economic changes are being registered, a fundamental critique of the capitalist economy that underlies such changes is missing.
3. Social factors (modes of regulation, habits of consumption, etc.) are rarely factored into the equation (apart from noting differences such as the auto dependency of North American cities versus the pedestrian nature of Hong Kong's mobility system).
4. Nature is seen as relatively static: material streams are described as mostly unchanging in character and itself not with a sense of agency but—in good engineering tradition—as an object of human ingenuity (see Keil & Boudreau, 2006, for an elaboration of this argument).

Much of this critical UPE discussion on metabolism has been taken up in a book edited by Nik Heynen, Maria Kaika, and Erik Swyngedouw (2006) titled *In the Nature of Cities: Urban Political Ecology and the Politics of Urban Metabolism*. The editors state programmatically: "Clearly, any materialist approach insists that 'nature' is an integral part of the 'metabolism' of social life. Social relations operate in and through metabolizing the 'natural' environment and transform both society and nature" (Heynen et al., 2006: 8). The significance of this statement in light of the foregoing fundamental capitalist critique is the acceptance of the embeddedness of all technical and symbolic aspects of sustainability (of metabolisms) in social relationships that govern the way physical and ecological relationships are established, sustained, and destroyed over time.

The work on metabolism flows into the emerging debate on "cyborg urbanization."[2] If metabolism is a biophysical process that serves as a model for the more complex and intertwined socionatural processes investigated under this heading in UPE, the "blurring of boundaries between the body and the city raises complexities in relation to our understanding of the human subject and the changing characteristics of human agency" (Gandy, 2005: 33). Gandy's comprehensive and critical overview article on the subject is a good guide to the intricate issues experienced in the cyborg city. Through overlapping agendas in the work of Agamben, Haraway, Law, Latour, and others, the debate on the cyborg city connects with the one on "post-humanism."[3] This then also relates to urbanization and emerging infectious disease (Ali & Keil, 2006; Gandy & Zumla, 2003) and the realization that urban political ecologies might soon find themselves in the center of a storm of pandemic proportions[4] (Davis, 2005).

The cyborg city is based also on "the idea of urban space as a prosthetic extension of the human body. . . . The cyborg metaphor not only reworks the metabolic preoccupations of the nineteenth-century industrial city but also extends to a contemporary body of ideas that we can term 'neo-organicist' on account of the deployment of biophysical metaphors for the interpretation of social and spatial complexity" (Gandy, 2005: 29). This neoorganicist view is "founded on the blurring of boundaries rather than their repeated delineation" (Gandy, 2005: 69). The city had been lifted out of the realm of the natural in the course of modernization processes in the 19th and 20th centuries and the scientific processes that established the basis for complex mechanical and technological networks such as water and sewer lines, which helped to quash disease that rested specifically and explicitly on the separation of germs from people, the organic from the societal, etc. This kind of denaturalization through continuous processes of ecological modernization and planning ingenuity is subject to our critique above. Used in this manner, the "cyborg city" reformulates this "neoorganicism" into a materialist urban theorem, which is relevant to any rethinking of sustainability: "The relational or dialectical cyborg, with its explicit engagement with different forms of capital—ranging from tangible manifestations to speculative abstractions" (Gandy, 2005: 37; see also Swyngedouw, 2004) affords the possibility of reformulating sustainability as a question of controlling capital at the center of urbanization.

The question is how a politics of sustainability could be engaged here to mobilize societal relationships with nature in order to stop capital— "the enemy of nature," as Kovel (2002) has it—from fueling the destruction of the world, human and natural. On the basis of the metabolic and cyborgian character of the city, of its material constitutionality, a politics of sustainability must include an agenda that redirects the devastating force of exchange value-oriented accumulation into a stream of use value-oriented products and services that help sustain human and natural metabolisms. Such a politics needs to resolve the challenges of the Latourian new constitution: "A cyborgian public realm, a la Latour, might, for example, include a whole variety of non-human organisms, yet-to-be-named assemblages of things and multitudinous chains of agency" (Gandy, 2005: 35). While recognizing the issue, Gandy explicitly warns against a full-blown inclusion of all potential actants into the polity of the city and points to the symbolic realm of the conceptions of the urban as the domain of progressive politics: "Part of the political challenge facing the hybrid city and its multifarious entanglements between the 'real' and the 'unreal' is to construct new kinds of autonomous spaces within which it is possible for different conceptions of the city to take shape" (2005: 41). As a consequence, politics needs to establish regulative processes that sustain use value-oriented socioecological relationships. At the center of these politics needs to be one issue: social justice. The final aspect of our urban political ecology of sustainability— together with the metabolic and cyborgian politics—is the politics of environmental justice.

Environmental justice is the direct antipode of the politics of ecological modernization, which is—in its hegemonic form—a politics of neoliberalization (Desfor & Keil, 2004). Environmental justice in urban environments cannot be universalized but is dependent on the circumstances of political struggle and socioeconomic relations. At the same time as its meaning is specific, it is also multiscalar: "The urbanization of environmental justice movements is part and parcel of the overall urbanization of life in this period of capitalist accumulation" (Debbané & Keil, 2004). But the purely material circumstances say little in themselves about the politics of sustainability that spring from such circumstances. Rather, "Injustice perceptions and justice demands are constructed through relative, scale-sensitive political and discursive processes" (Debbané & Keil, 2004; Keil & Debbané, 2005). In political terms, urban environmental justice is constructed in a forcefield of racialization and spatialization (Debbané & Keil, 2004) and as long as a politics of sustainability remains silent on these issues. This point needs no elaboration here. It has been forcefully made by Agyeman, Bullard, and Evans (2003), who have posed the challenge of justice to the ongoing quest for environmental development. What needs to be added here, though, is the impression that, as Andrew Dobson has said, "reds and greens have fundamentally different objectives" (2003: 83) needs to be overcome. This is not, as Dobson would have it, because these are mostly strategic, fundamental political differences that seem entirely incompatible. It is because the realities through which justice and sustainability are constructed are necessarily intertwined more than separate: if there is a reason to sustain the natural environments of the globe, it is because they hold the promise of a more equitable world. One makes no sense without the other.

CONCLUSION

I have, in this chapter, provided a fundamental critique of approaches of sustainability and ecological modernization. Their genesis as terms of international politics has been shown to be connected intrinsically to the demise of the Fordist–Keynesian regime of accumulation and to the emergence of a globalized neoliberal, post-Fordist regime under American hegemony. The tradition of sounding the warning bells about the planet's ecological health that started with the politics of Rachel Carson and the Club of Rome and ended, for now, in the Millennium Ecosystem Assessment provides us with a clearer sense of the world ecological problematique. Unfortunately, the urgency of these warnings has not translated into a similarly incisive world resolutique. This is partly due to the limits of these interventions themselves (which operate inside the aggressive expansionary capitalism we currently live in) and it is partly to be credited, as Buell shows so clearly, to the politics of the apocalypse, which has tired as a strategy of social and environmental change. I have suggested in this chapter that a critical urban political ecology may be a possible way to reconstruct both a critical socioecological politics and a less-than-apocalyptic concrete politics of systemic change. The ideas of metabolism and cyborg urbanism were introduced as major conduits of such a politics. At the end of this analysis, a classical "what is to be done" question remains to be answered. If the suggestion is correct that urban political ecology is an arena of change, there are two major strategic areas in which this change can be envisioned at this point in time:

1. The existing mainstream environmental urbanist propositions (new urbanism, etc.) will need to be politicized at the seemingly opposed ends of social justice and ecological change.

2. The emerging areas of UPE must be radicalized consistently to free them from their shackles of human-centered analysis and middle-class politics. They must be redefined as "posthumanist" projects that allow nonhuman actants carefully into the democratic process, and they must be uncompromisingly directed toward the pursuit of eradicating societal injustices that stand in the way of sustainability.

NOTES

1. Similarly, Dennis and Donnella Meadows, the principal authors of *The Limits to Growth* (Meadows, Meadows, Randers, & Behrens, 1972), refined their analysis during the 1990s. Working with more sophisticated computer models, they now introduced the important distinction between "growth" and "development": " 'to grow' means to increase in size by the assimilation or accretion of materials. 'to develop' means to expand or realize the potentialities of; to bring to a fuller, greater, or better state" (cited in Becker & Jahn, 1998: 76). In their critique of the Meadows's newer work Becker and Jahn point to their latent Malthusian tendencies and lack of understanding of destructive tendencies.

2. Franz Hartmann must be credited with being one of the first users of this notion in the context of his dissertation on urban political ecology (1998); see also Swyngedouw (1996, 2004).

3. See the exchange on "mapping posthumanism" in *Environment and Planning A*, vol. 36 (2004), pp. 1341–1363, with Noel Castree, Catherine Nash, Neil Badmington, Bruce Braun, Jonathan Murdoch, and Sarah Whatmore. Bruce Braun, in particular, has started to work on the city as a " 'more-than-human' assemblage," using the concept of the "biopolitical city" (Braun, 2005).

4. We know from Eric Klinenberg's (2002) work on the Chicago heat wave the potentially catastrophic relationships of urban social ecologies, the built environment, and life and death.

REFERENCES

Agyeman, J., Bullard, R. D., & Evans, B. (Eds.). (2003). *Just sustainabilities: Development in an unequal world*. Cambridge, MA: MIT Press.

Ali, S. H., & Keil, R. (2006). Global cities and the spread of infectious disease: The case of severe acute respiratory syndrome in Toronto, Canada. *Urban Studies, 43*(3), 491–510.

Alvater, E. (1993). *The future of the market*. London: Verso.

Barber, B. (1995). *Jihad vs. McWorld*. New York: Times Books.

Becker, E., & Jahn, T. (1998). Growth or development? In R. Keil, D. J. Bell, P. Penz, &

L. Fawcett (Eds.), *Political ecology: Global and local* (pp. 68–83). London: Routledge.

Bond, P. (2000). Economic growth, ecological modernization or environmental justice?: Conflicting discourses in post-apartheid South Africa. *Capitalism, Nature, Socialism, 11*(1), 33–61.

Bond, P. (2002). *Unsustainable South Africa.* London: Merlin Press.

Braun, B. (2005, April 29). *SARS and the posthuman city.* Paper presented at the SARS and the Global City Workshop, York University. Available at *http://www.yorku. ca/sars2003.*

Buell, F. (2003). *From apocalypse to way of life: Environmental crisis in the American century.* London: Routledge.

Davis, M. (2006). *Planet of slums.* London: Verso.

Debbané, A. M., & Keil, R. (2004). Multiple disconnections: Environmental justice, urban sustainability and water. *Space and Polity, 8*(2), 209–225.

Desfor, G., & Keil, R. (2004). *Nature and the city: Making environmental policy in Toronto and Los Angeles.* Tucson: University of Arizona Press.

Dobson, A. (2003). Social justice and environmental sustainability: Ne'er the twain shall meet. In J. Agyeman, R. D. Bullard, B. Evans (Eds.), *Just sustainabilities: Development in an unequal world* (pp. 83–98). Cambridge, MA: MIT Press.

Gandy, M. (2005). Cyborg urbanization: Complexity and monstrosity in the contemporary city. *International Journal of Urban and Regional Research, 29*(1), 26–49.

Gandy, M., & Zumla, A. (Eds.). (2003). *The return of the white plague: Global poverty and the "new" tuberculosis.* London: Verso.

Girardet, H. (1992). *The Gaia atlas of cities.* New York: Anchor Books.

Girardet, H. (1999). *Creating sustainable cities.* Totnes, UK: Green Books.

Görg, C. (2003). *Regulation der Naturverhältnisse: Zu einer kritischen theorie der ökologischen krise.* Munster: Westfalisches Dampfboot.

Grenfell, D. (2006). Beyond the local and the global: Scales of resistance, repression, and sustainability. In J. Johnston, M. Gismondi, & J. Goodman (Eds.), *Nature's revenge: Reclaiming sustainability in an age of globalization* (pp. 225–244). Peterborough, ON: Broadview Press.

Hall, P. (2005). *The sustainable city: A mythical beast.* L'Enfant Lecture on City Planning and Design, National Building Museum and the American Planning Association, Washington, DC. Transcript obtainable at *http://www.nbm.org/Events/Calendar/SirPeterHall NBMLenfantLecture.pdf* (last accessed on April 2, 2007).

Hajer, M. (2003). Policy without polity? Policy analysis and the institutional void. *Policy Sciences, 36,* 175–195.

Hardt, M., & Negri, T. (2000). *Empire.* Cambridge, MA: Harvard University Press.

Hart, M., & Negri, T. (2004). *Multitude.* Cambridge, MA: Penguin.

Hartmann, F. (1998). *Nature in the city: Urban ecological politics in Toronto.* Unpublished PhD thesis, Department of Political Science, York University, Toronto.

Harvey, D. (1996). *Justice, nature, and the geography of difference.* Oxford: Blackwell.

Hays, S. P. (1987). *Beauty, health and permanence: Environmental politics in the United States, 1955–1985.* New York: Cambridge University Press.

Heynen, N. (2003). The scalar production of injustice within the urban forest. *Antipode, 35*(5), 980–998.

Heynen, N., Kaika, M., & Swyngedouw, E. (Eds.). (2006). *In the nature of cities: Urban political ecology and the politics of urban metabolism.* London: Routledge.

Johnston, J., Gismondi, M., & Goodman, J. (2006b). Politicizing exhaustion: Ecosocial crisis and the geographic challenge for cosmopolitans. In J. Johnston, M. Gismondi, & J. Goodman (Eds.), *Nature's revenge: Reclaiming sustainability in an age of globalization* (pp. 13–36). Peterborough, ON: Broadview Press.

Keil, R., & Boudreau, J.-A. (2006). Metropolitics and metabolics: Rolling out environmentalism in Toronto. In N. Heynen, M. Kaika, & E. Swyngedouw (Eds.), *In the nature of cities: Urban political ecology and the politics of urban metabolism*. London: Routledge.

Keil, R., & Debbané, A.-M. (2005). Scaling discourse analysis: Experiences from Hermanus, South Africa and Walvis Bay, Namibia. *Journal of Environmental Policy and Planning, 7*(3), 257–276.

Klinenberg, E. (2002). *Heatwave: A social autopsy of disaster in Chicago.* Chicago: University of Chicago Press.

Kovel, J. (2002). *The enemy of nature: The end of capitalism or the end of the world?* Nova Scotia: Fernwood.

Krueger, R., & Agyeman, J. (in press). Sustainability schizophrenia or "actually existing sustainabilities": The politics and promise of a sustainability agenda in the U.S. *Geoforum.*

Latour, B. (2004). *Politics of nature: How to bring the sciences into democracy.* Cambridge, MA, and London: Harvard University Press.

Latour, B. (2005). *Reassembling the social: An introduction to actor–network–theory.* Oxford, UK: Oxford University Press.

Lipietz, A. (1992). *Towards a new economic order: Postfordism, ecology and democracy.* New York: Oxford University Press.

Low, N., Gleeson, B., Elander, I., & Lidskog, R. (2000). *Consuming cities: The urban environment in the global economy after the Rio Declaration.* New York: Routledge.

Meadows, D., Meadows, H., Randers, D. L., & Behrens, J. W. III. (1972). *The limits to growth.* New York: Universe Books.

Millennium Ecosystem Assessment. (2005, March). *Living beyond our means: Natural assets and human well-being.* Statement from the Board. (Available at *http://ma.caudillweb.com/eu/Products.aspx.*)

Neuwirth, R. (2005). *Shadow cities: A billion squatters, a new urban world.* New York: Routledge.

Newcome, K., Kalma, J., & Aston, A. (1978). The metabolism of a city: The case of Hong Kong, *Ambio, 7*(1), 3–15.

O'Connor, J. (1998). *Natural causes.* New York: Guilford.

Rogers, R., & Gumuchdjian, P. (1997). *Cities for a small planet.* London: Faber & Faber.

Rogers, R., & Power, A. (2000). *Cities for a small country.* London: Faber & Faber.

Sahely, H., Dudding, S., & Kennedy, C. (2003). Estimating the urban metabolism of Canadian cities: Greater Toronto area case study. *Canadian Journal of Civil Engineering, 30*, 468–483.

Sparke, M. (2005). *The space of theory: Postfoundational geographies of the nation-state.* Minneapolis and London: University of Minnesota Press.

Sywngedouw, E. (1996). The city as a hybrid: On nature, society and cyborg urbanization. *Capitalism, Nature, Socialism, 7*(2), 65–80.

Swyngedouw, E. (2004). *Social power and the urbanization of water: Flows of power.* Oxford, UK: Oxford University Press.

Swyngedouw, E., & Heynen, N. (2003). Urban political ecology, justice and the political ecology of scale. *Antipode, 35*(5), 898–918.

Wackernagel, I., & Rees, W. E. (1996). *Our ecological footprint: Reducing human impact on the earth.* Gabriola Island, BC, and Philadelphia: New Society Publishers.

Waren-Rhodes, K., & Koenig, A. (2001). Escalating trends in the urban metabolism of Hong Kong: 1971–1997. *Ambio, 30*(7), 429–438.

While, A., Jonas, A., & Gibbs, D. (2004, September). The environment and the entrepreneurial city: Searching for the urban "sustainability fix" in Manchester and Leeds. *International Journal of Urban and Regional Research, 28*(3), 549–569.

Wolman, A. (1965). The metabolism of cities. *Scientific American, 213*(3), 178–193.

World Commission on Environment and Development. (1987). *Our common future.* Oxford, UK: Oxford University Press.

York, R., & Rosa, E. A. (2003, September). Key challenges to ecological modernization theory: Institutional efficacy, case study evidence, units of analysis. *Organization and Environment, 16*(3), 273–288.

CHAPTER 3

Microgeographies and Microruptures

The Politics of Gender in the Theory and Practice of Sustainability

SUSAN BUCKINGHAM

In a piece written when the concept of sustainability was in its infancy—when people were still questioning its semantic value and it was far from being a political commonplace, at least in Europe—Visvanathan eloquently dismissed the utility of the term "sustainable development": "sustainability and development belong to different, almost incommensurable worlds. We were told in catechism class that even God cannot square the circle. Sustainable development is another example of a similar exercise" (Visvanathan, 1991: 238).

I would add, and intend to use this chapter to justify why, that sustainability is especially problematic once it is coupled with "gender," or, indeed, any other signifier of relative disadvantage, for one thing society definitely does not need is for gender relations to be sustained in the unequal forms in which they currently exist. Much (although I would argue, not enough) has been written about the relationship between gender and environment, and much of this literature explains how contemporary gender relations expose women, more than men, to environmental problems. (For an overview of this body of work, see Buckingham-Hatfield, 2000; Buckingham, Budd, Lynn, Murphy, & Sutton, 2005; Mellor, 1992; Mies & Shiva, 1993.) It also argues that a social system that creates and supports gender inequality through continuing to privilege masculine qualities such as aggression and competition has significant negative impacts on the environment. Sherilyn MacGregor has reached similar conclusions in her contribution to a recent book on environmental citizenship in which she argues that environmental sustainability is impossible without gender equality (MacGregor, 2006).

While it is possible to read an essentialist argument into some of the earlier work on feminist ecology (see, e.g., Merchant, 1996), most of the more recent literature, and particularly that on which this chapter draws, stresses the importance of prevailing social structures that continue to benefit masculine social elites in influencing attitudes toward the environment. It is the circularity of these structures, in which prevailing social hierarchies are maintained by the self-interests of their beneficiaries, with which this chapter is concerned. In particular it is concerned with the mechanisms that have the potential to disrupt this circularity.

If anything can be rescued from the term "sustainability" it is the recognition, at least, that the economic, environmental, and social are inextricably bound. Elsewhere I have argued with my coauthors that socioeconomic structures across the world work against gender equality, and yet we concluded optimistically that women seize what power and control they have over their lives despite these structures (Buckingham & Lievesley, 2006). In arguing this, I found Alain Lipietz's concept of "microruptures" powerful in creating the potential for change (Lipietz, 2000) and will suggest here that there are points at which existing policy and practice can be ruptured to make space for change in which environmental issues and gender equality can be addressed in mutually constructive ways. But this is not going to be easy, as gender inequalities permeate—and construct—the environmental movement, as well as government policymaking and business practice.

The second concept that will be used to analyze the neglect of gender in sustainability discourses is that of microgeographies. While poverty, race, and ethnicity are strongly contoured, gender (and consequently gender–environment) relations are often hidden within bodies and households (e.g., see Butler, 1990). Feminist theorists have long argued that this invisibility has worked against theoretical and political development of gender equality (Lister, 1997), and this is equally true of gender sensitive environmental justice analysis and programs. Swynge-douw and Heynan (2003: 913) in their political ecology review of environmental justice note, though not in specific reference to gender, that the scalar capacities of social groups are reflected in the wider scalar hierarchy in which broader, larger scales are accorded more importance than the smallest. In Marston's (2000) review of the social constructions of scale, she accuses Peter Taylor of focusing on the "world economy" as the scale that "really matters" (Marston, 2000: 226). The "urban scale" experience is where capital accumulation that is ultimately organized at the global scale materializes (but no mention of the neighborhood, household, or bodily scale). Because women have the greatest traction at the smaller scales (the household and neighborhood), it stands to reason that any changes—microruptures—they might achieve will be at this level. However, the ability to move across scales is predicated on the ability of oppositional movements to "take advantages of resources at one scale to overcome the constraints encountered at different scales" (Staeheli, 1994: 388). This, Staeheli argues, defines the power of actors and their "potential for pressing their claims" (p. 388). This implies that there are two changes that are needed to gender-democratize sustainability policy and practice. Scaling up, more women need to be involved in high-level decision making. Given that one of the arguments for more women to participate is to recognize and value the diversity of experience with regard to the environment in this decision making, it is axiomatic that the increase in the number of women needs to reflect the diversity of women's experience. Scaling down requires an increase in the recognition of what local decision making can achieve, which must involve a genuine devolution of decision making to the locality and community.

In order to pursue these arguments I first consider the shape of the status quo in terms of the gendering of government, business, and the environmental movement, followed by the continuing economic disadvantages that reinforce the feminization of poverty, and consequently makes gender an environmental justice issue, albeit one to which the environmental justice movement gives scant prominence. Building on this, I will then argue that there are organizations that recognize that environmental sustainability requires social equity to be achieved, and I will particularly examine the proposal that "gender mainstreaming," as advocated by the United Nations, World Bank, and European Union, has the potential to address broad and interlinked issues concerning sustainability if the barriers set up by national governments, business/industry, and the prevailing environmental campaigning culture can be overcome. To some degree this illustrates the ability of women's campaigning groups to "jump scale" in the 1990s, taking advantage of the relative openness of international organizations when national legislatures appeared resistant to change. (See, for example, Marston's analysis of the U.S. antinuclear weapons movement, which was able to transcend local and national state political opportunity to direct their protests to global businesses headquartered in Massachusetts ([Marston, 2000: 224]). In particular I will draw on research undertaken for the European Union in 2003 that examined the scope of gender mainstreaming in municipal waste management. The findings of this, together with the following review of research elsewhere, confirm that changing, rather than sustaining or reinforcing, gender roles is critical if environmental problems currently facing the world are to be overcome equitably.

DEFINING APPROACHES
AND PROBLEMATIZING SUSTAINABILITY

Sustainability

There has been much written about sustainability in terms of sustaining an environment that will serve future generations as effectively as it serves ours, but beyond this there is much disagreement about what, and how meaningful or useful a term, "sustainability" actually is. We might question whether sustainability is about sustaining the current environment (i.e., not allowing it to worsen) or about enabling it to be sustained for future generations (i.e., not to damage it beyond its capacity for reversal). The first is effectively the position of climate change agreements, which require a stabilization of greenhouse gas emissions at 1990 levels, whereas the second can be illustrated by agreements on biodiversity that require a sufficient species critical mass to be preserved for future utilization. Alternatively it can be about enabling communities to sustain themselves or be sustained, a particular thrust of the U.K. government's "Sustainable Communities" program, which defines a "sustainable community" as a place where "people want to live and work, now and in the future. They meet the diverse needs of existing and future residents, are sensitive to their environment, and contribute to a high quality of life. They are safe and inclusive, well planned, built and run, and offer equality of opportunity and good services for all" (Department for Environment, Food and Rural Affairs, 2005: 121).

"Sustainability" and "sustainable development" are political concepts, although little of the research on these concepts theorizes underlying power relations (Lipietz, 1995). These power relations (including patriarchy), and the articulation of sustainability with the production—and reproduction—of capital, place it preeminently in the intellectual domain of political ecology. So, although I think "sustainability" is an unhelpful way of envisioning an environmental and social future in that it makes assumptions that what we have at present is something we want to capture or secure, it does provide a conceptually useful toehold for analyzing the power relations between human beings and nature that is political ecology. This is an important distinction to make, as sustainability is particularly problematic when we consider gender relations, which, many would argue, need a complete overhaul.

Gender and Environment Relations

Gender relations are the neglected social dynamic of "sustainability"—as they have been with every other revolution that has sought one dimension of social change, on the back of continuing uneven gender relations. Political revolutions the world over have generated support from women on the basis that once the class war has been won, then gender inequalities can begin to be addressed (see Hunt on justifying the "war on terror" in Afghanistan and Iraq, 2006; Lievesley on Cuba, 2004, and on Latin America and Russia, 2006a, 2006b; Mehdid on Algeria, 1996). As these authors demonstrate, once relegated in this way, gender inequalities are never redressed. The reason for this is that the solution to the class/colonial inequalities that these revolutions promote is not only unequal to the task of creating gender equality—more than that, it is born of and grounded in these inequalities.

Andrea Nightingale, in a poststructuralist analysis of the relationship between gender and environment, argues that so long as "environment" remains the focus for attention, as it appears to be within mainstream environmental debates, then, in the face of impending "risk," "it is considered difficult to make a clear argument about why we need to care if men and women have different experiences and knowledge of that risk" (Nightingale, 2006: 170). Through her work in Nepal's community forests, Nightingale proposes that "gender" and "environment" are mutually constituted and that attention to the ways in which gender is "performed" "is crucial for understanding how environmental issues come to be environmental in the first place" (Nightingale, 2006: 172).

While I am more inclined to see gender-environment relations more as a product of powerful and endemic structural relations, Nightingale's argument supports my own in that the strategies being developed to *sustain* our current environment rest upon gender inequalities and gender roles. I will argue that contemporary notions of sustainability, like the political revolutions in the past, rest on presumptions of inequality. Lipietz has argued, in support of political ecology, that "our relationship with nature is bad, because relationships between human beings are already bad" (1995: 148). To take this point further, the nature of these relationships must then be examined, and gender inequality is a persistent factor in this.

Microruptures

Alain Lipietz has argued the potency of microruptures as a way of achieving incremental change, from a position of "radical democratism" and as a "go-between from social movements to institutions" (2000). Coming from a Marxist intellectual tradition and a position in the French Green party, he argues that political ecology is the "21st century inheritor of Marxism, in that it is the only viable response to continuing problems." In so doing, Lipietz claims both an analytical/theoretical and practical role for political ecology in society–environmental relations, which he considers in need of revision. His experience suggests the potential for some positive change to be achieved within and despite broader social and economic structures. In an earlier polemic, he argues against the charge of "why try to do something when there are billions of people around us conspiring against our environment? And why try to do something when what one achieves is tiny in comparison with what remains to be done?", concluding "everybody's environment is every one of us . . . it is worth the effort, and . . . there are billions of us" (Lipietz, 1995: 151). Following Lipietz, political ecology, then, calls for both theoretical analysis of the political and social processes involved in the production of "sustainability" and practical action, which recruits "sustainability" for social change.

THE STATUS QUO

Decision making is structured by inequalities, and this affects decisions made in the environmental sector as much as anywhere else: in government, industry, and also in the campaigning field. This section briefly examines these three broad areas to reveal the persistence of gender inequality and the potential impacts of this on sustainability.

Government

Women constitute around 25% of Members of the European Parliament (MEPs) and 18% of U.K. Members of Parliament—with around 30% of the U.K. Cabinet being women. Table 3.1 shows the low participation of women in formal political activity in the European Union, while Table 3.2 illustrates a similar pattern elsewhere. In the United States in 2005, 14% of the 100 U.S. senators and 16% of state governors were women. Indeed, it is notable from the selection in Table 3.2 that a continuing rise in the number of women in parliaments is by no means an inexorable trend, with sizable declines recorded in a number of countries in both the global South and in countries previously under communist control in central and eastern Europe.

Worldwide there are few political fora where women are able to form a critical mass, which Bhattar (2001) argues is necessary for women to support one another in policy initiatives, to be a catalyst for other women to become involved, and to be in a position to allocate and control resources. About 30–35% is the proportion that generally is considered to achieve this critical mass. Of course, such a proportion in itself is only a proxy for the degree of change—simply having women in power is no guarantee that attitudes toward anything, and particularly gender equality or environmental policy, will change—but it does represent the potential for change, which, as will be illustrated below, can make a difference. Although it is not possible to establish a causal link, it is worth noting that in countries with relatively high proportions of

TABLE 3.1. Positions Held by Women and Men in European Union Institutions, 2004

Role/institution	% women	% men
Members of European Parliament	28	72
Senior Ministerial Positions—EU Average	25	75
Junior Ministerial Positions—EU Average	22	78
Members of Parliament—EU Average	23	77
Members of Upper House—EU Average	21	79

Note. Data from European Commission, Employment and Social Affairs (2004).

TABLE 3.2. Women in Politics in Selected Countries

Country	% of women-held seats in national parliaments in 1990	% of women-held seats in national parliaments in 2003
Australia	6.1	27.1
Canada	13.3	26.5
Bangladesh	10.3	2.0
Dominica	10.0	18.8
Guyana	36.9	20.0
Hungary	20.7	9.8
India	5.0	9.6
Romania	34.4	8.2
Seychelles	16.0	29.4
South Africa	2.8	30.7
Sweden	38.4	45.3
Uganda	12.2	24.7

Note. Data from United Nations (2005).

women as legislators there is also a greater emphasis on sustainability, as analyses of Welsh, Swedish, and Norwegian policy confirms. Sweden was the first country worldwide to introduce a quota system to increase the number of women in Parliament and was the first country to reach something approaching gender parity in their representation. In the devolved government in Wales, 56% of its Cabinet appointments were women in 2005; it was also the only U.K. legislature that had a statutory "sustainable development" policy. A Commonwealth Secretariat document suggests that "even a few women in the corridors of power lead to a more participatory, less autocratic style of government" (Commonwealth Secretariat, 1998), and it is, therefore, tempting to conjecture that such a critical mass of women can instigate enough microruptures to effect significant change. It is, perhaps, important to stress here that there are many reasons why women are likely to bring a different attitude toward the environment to decision making, which a later section on social and economic disadvantage will develop.

National power structures are generally characterized by an inverse relationship between degrees of power, on the one hand, and degrees of localness and higher proportions of women, on the other. Women are generally more prominent in local government—in England and Wales, for example, around 30% of local councillors are women—notably women are most active in grassroots community action, which is often the only forum in which women feel they can express their concerns, substantiating Swyngedouw and Heynan's earlier point about social hierarchies reflecting scalar hierarchies.

Business and Industry

Business and industrial sectors worldwide have always been overwhelmingly dominated by men and by masculine ways of operating. This is evidenced by the pitifully few women who sit on the boards of major companies or who hold chief executive posts. For example, only 8% of boardroom seats on Europe's top-listed 200 companies were held by women in 2004 (European BoardWomen Monitor, 2004) . The most recently recorded *Guardian* survey of boardroom demographics of the FTSE 100 (the 100 largest companies on the Financial Times Stock Exchange) reveals a downturn: compared to 20 woman executive directors in 2005 (vs. 15 and 17 in 2003 and 2004, respectively), in 2006 there were just 12, with only 112 nonexecutive directors (compared to 122 in 2005). Twenty-seven of the 100 companies had no female board members whatsoever (*The Guardian*, 2006). Furthermore, a Deloitte & Touche report reports "huge gender imbalance in the boardroom. There has been no increase in the number of female executive board members, and only a 1% increase in the number of female non-executive directors. Women only make up 3% of executive directors and 10% of non-executive directors across the FTSE 350" (Deloitte & Touche LLP, 2006).

There is also a conspicuous lack of women in key fields that make a significant impact on, or potential contribution to, the environment: energy, transport, water, waste management, and building. For example, in Germany, 20% of employees in the energy industry are women, and only 3% of management and professional staff are women (Climate for Change, 2002). The professions that support these industries are likewise gendered: planning, architecture, surveying, engineering, physics, chemistry. For example, in 1994, Greed found that less than 10% of practicing architects, less than 20% of accredited planners, and less than 10% of surveyors in the United Kingdom were women. Although there are campaigns to encourage more young women to enter these professions, there continues to be a significant gender imbalance. Inequalities in government and business are well known and discussed. However, very little analysis has been made of the environmental movement itself.

Environmental Campaigning Groups

The majority of environmental campaigning groups tend to replicate the unequal gender relations found elsewhere in industry, business, and government that Joni Seager's review of the North American Environmental Movement revealed in the early 1990s (Seager, 1993). In a review of environmental nongovernmental organizations (ENGOs) that are members of the European Union's Civil Society Contact Group, 43% of the heads of these groups were found to be women, while only 24% of members of their highest decision-making bodies were women. Contact Group members have agreed as a condition of their membership to promote gender parity, so it is reasonable to assume that ENGOs that are not part of this group are unlikely to have better women's representation in their senior management. Table 3.3 gives a good indication of the gender profile of a number of the largest British environmental organizations in the financial year 2004/2005.

Table 3.3 shows that governance of these major environmental campaigning organizations, which are increasingly significant players in determining governmental policies, is markedly gendered. Only one of the six organizations had a female chief executive, and only one had a female chair of the board of trustees, and all the boards of trustees were predominantly male. Further reading of the annual reports from which these data are drawn reveals that councils, regional boards, and so on are similarly gendered. While it is not possible to conclude with certainty that this has affected the groups' policies and campaigns, given the weight of evidence in other organizations suggesting that the gender of decision makers affects the nature of decisions taken (see Bhattar, 2001, referred to earlier), it would be highly unusual if this were not the case in environmental campaigning.

The founders of the Women's Environmental Network (WEN) in the United Kingdom in 1988 cited as one of their motivations the fact that other ENGOs at the time failed to address women's environmental concerns, or the background underpinning these, a situation largely unchanged during the ensuing years (Women's Environmental Network, 1993). That WEN faces persistent funding difficulties is largely explained by a recent report undertaken by the Women's Resource Centre, which found women's campaigning groups in the United Kingdom to be

TABLE 3.3. Gender Profile of Major Environmental U.K. Organizations, 2004–2005.

Organization revenues	CEO	Chair	Board of trustees
Council for the Protection of Rural England	Male	Male	Male president, all 5 vice presidents male; all 5 national executives male
Friends of the Earth	Male	Male	9 male, 4 female
Greenpeace	Male	Female	All male
National Trust	Female	Male	9 male, 2 female
Royal Society for the Protection of Birds	Male	Male	Key positions male
WWF U.K.	Male	Male	11 male, 2 female

Note. Data from annual reports.

considerably underresourced relative to others. While charities working specifically with women constitute 7% of all charities registered with the Charities Commission, they command only 1.2% of funding (Women's Resource Centre, 2006). A number of reasons are given for this relative paucity of funding, including a shift of government awards to procurement and tendering, which requires the voluntary sector to "frame funding applications within 'the paradigm of need recognised by funders' " (WRC, 2006: 10–13). The WRC report claims that "political rhetoric about women's equality fails to be matched by public investment in women's organisation" (WRC, 2006: 63), and cites evidence such as the Equal Opportunities Commission's receiving less funding than any of the other equalities commissions (WRC, 2006: 63).

Given the observations of Neil Carter (2001) and Chris Rootes (1999) on how campaigning groups become more "incorporated" over time, the gendered nature of their own structures mimicking those in the governments and industries they lobby is highly likely to influence the issues raised and the proposed strategies for redressing them. If, as Nightingale suggests, gender structures environmental issues, then campaigning issues identified by organizations heavily male-dominated at the senior level are equally likely to be gendered.

CONTINUING ECONOMIC AND SOCIAL DISADVANTAGE

Persistent income disparities reinforce the feminization of poverty and consequently make gender an environmental justice issue, although the environmental justice movement fails, I would argue, to address the gendered nature of poverty and race sufficiently to do it justice.

A survey of poverty and social exclusion published by the U.K. Equal Opportunities Commission (EOC) reported that 36% of women, compared to 30% of men, lived in households with incomes less than 60% of the median. In addition, women were more likely to be poor on all four dimensions of poverty used by the U.K. government (Bradshaw, Finch, Kemp, Mayhew, & Williams, 2003). Even when controlling for other factors such as labor market status, number and age of children, household composition, and age, there was still a clear gender dimension to poverty. In addition, women who are single pensioners, unemployed, of Pakistani or Bangladeshi origin, a teenage householder, or tenant, are *more* likely to be poor than men with the *same* characteristics. Those likely to experience the greatest degree of poverty are lone mothers and older single women.

Women in full-time paid work in the United Kingdom will, on average, earn 81% of the hourly wages earned by men in full-time work. Although this gap is currently closing, the hourly pay-rate gap between women in part-time work and men in full-time work is widening such that these women earn only 61% of the hourly rate of men in full-time work (Kingsmill, 2001). The EOC report also suggests that there is unequal poverty within the household, with some women having unequal access to household earnings in cases where the male partner is the main earner, and that mothers sometimes forgo consumption in order to meet the needs of the rest of their family. In the United States, the situation is even more polarized. The National Commission on Pay Equity has reported that women working full-time yearround earned 76% of the equivalent male wage. Compared with this same earnings figure for men of all races, African American and Hispanic women earned even smaller proportions (60% and 55%, respectively) (National Commission on Pay Equity, 2004).

Links between poverty and poor environmental quality are well documented (FoE Scotland, 2000), and the foregoing discussion suggests that women are on average, consequently, more likely (because they are women) to experience poorer environmental quality than men. Those with low income will be exposed to higher rates of traffic and industrial pollution, as they do not have the resources to buy themselves out of the most environmentally degraded areas. They are more likely to experience fuel and food poverty by living in drafty, poorly insulated, and damp accommodations, and to be more malnourished as a result of being less discriminating with regard to food quality. U.S. Department of Agriculture data for 2002 established that two-thirds of all households that are "food insecure" (i.e., with limited or uncertain availability of nutritionally adequate foods) are headed by a single parent, roughly 90% of whom are women (National Anti-Hunger Organizations, 2004).

Worldwide, around 1.2 billion people currently live below the UN-defined poverty line of $1 a day, while a further 2.8 billion earn less than $2 a day. Some 70% of these people in poverty are women (Dankleman, 2002). The education gap between girls and boys in the developing world is still sufficiently wide to ensure that gender inequalities will persist into the next adult generation, with UNESCO reporting that girl–boy secondary school enrollment ratios are standing at 0.96 in North Africa and the Middle East, 0.77 in sub-Saharan Africa, 0.95 in East Asia and the Pacific, and 0.86 in South Asia. Only in Latin America (1.07, but a decline since a high 1.14 in 1990) is this reversed (United Nations Millennium Project, 2005). Women constituted 50% or more of nonagricultural wage employment in only 17 of 110 countries reporting data, with 25–49% in a further 76. These figures suggest a lack of independent income for more than half—rising to around 80%—of all women in some global regions. These concerns are articulated as some of the UN Millennium goals, which are far from being met, and call into question the effectiveness of international legislation and agreements.

Given this overrepresentation of women in communities in poverty, it is not difficult to imagine a gendered geography of environmental injustice. However, the nature of gender relations, manifested in the microgeographies of households, interpersonal relationships, and the body, generates more complicated, layered, and less obviously visible patterns of inequality than those evident with race, ethnicity, and poverty. These microgeographies are, arguably, one reason why gender is not considered an important component of sustainability, as the research on waste management will later show.

Structural factors that determine women's economic disadvantage also influence the roles they play in the household. The household is increasingly being seen as a site of consumption, and nowhere is this so evident as in green marketing. Yet, as MacGregor (2006: 110) notes, "The question of how green practices in the private sphere are to be initiated, distributed and sustained is seldom, if ever, asked." MacGregor argues that the increasing emphasis on citizens' "responsibilities" with regard to the environment have a particular impact on women in their role as both domestic worker and carer. The framing of the citizen as a "responsible consumer" requires mindful purchasing and disposal habits that produce a "paradoxical coupling of labour- and time-intensive green lifestyle changes with increased active participation in the public sphere" (2006: 102). Although the discourse of green rights and responsibilities assumes a gender neutrality, in reality this "masks realities and specificities of gender inequality" and "reveals a lack of consideration for the politics of the private sphere" (pp. 106–107). Indeed, European research on the gendered dimension of municipal waste disposal, which will be discussed more fully later in the chapter, revealed the extent to which recycling and other green forms of waste disposal and minimization were largely the province of women in the household. Successive surveys of environmental attitudes reveal that involvement in the quotidian tasks needed to reduce environmental impact is routinely higher for women compared to men (e.g., 54% of women take paper for recycling, compared with 52% of men; 13% of women buy goods with less packaging compared with 11% men; 19% of women bought organic food, compared with 17% of men [Department for Environment, Food and Rural Affairs, 2002]).

Since gender is inscribed in one's relationship to work and environment, it is critical that all the emerging inequalities be considered commensurately. For example, when "environmental" problems are considered to be more important than social inequalities, strategies taken to address these problems can make these social inequalities worse. Nightingale (2006) has noted in Nepal how the social practices of leaf litter collection are gendered; since these practices are now being seen as ecologically destructive, this has led to changes in practice. However, without the involvement of women in this negotiation, or the involvement of men in the collection, the revised practice necessitates women being involved in physically and temporally more intensive work, which in turn reinforces gender inequality. Changing ecological practice without addressing gendered practices, then, can be problematic (Nightingale, 2006: 176–177).

In the West there are parallels regarding proposals for ways of reducing environmental damage that, without addressing prevailing gender relations, can result in considerably extended burdens for women. The following example illustrates how gender relations in the home and in the work economy, together with government transport, work, and education policy, combine to produce a particular environmental problem. The various campaigns in the United Kingdom to reduce the car traffic involved in the "school run," where 40% of primary school children are now driven to school (Department for Transport, 2005), focus on the fact that around 20% of morning traffic congestion is caused by the school run, in which women are depicted as the key culprits. Now, while it is undoubtedly true that the majority of school run chauffeurs are women (although not the majority of drivers overall in the United Kingdom), this is a function of the fact that, despite women's constituting 49% of all employees in the United Kingdom (and 38% of all full-time workers; Equal Opportunities Commission, 2006) the latest data available indicate that they still undertake the overwhelming majority of domestic and dependent caring roles (Her Majesty's Government, 1999). This often necessitates "trip chaining" the school run with grocery shopping and paid work and may be the only way in which women can effectively combine their responsibilities in the time available. Indeed, there is an argument to suggest that wider availability of the car has increasingly tied women to these social roles (Law, 1999; Dowling, 2000; Barker, 2006). While accusing these chauffeurs of contributing to congestion and global climate change, there is a disingenuous lack of attention to government decisions over the years to concentrate education in larger schools, or to offer greater locational choice in schools to families who have the resources to transport their children, to restrict school buses for collective pupil transport, or to clear congestion generated by other means to enable children to be able to walk or cycle more safely to school. The situation also reflects a general lack of flexibility regarding work practices, despite EU initiatives relating to worklife–homelife balance, and an economic situation in at least the United Kingdom and the United States, where in many areas families can only survive if both parents maximize their earning capacity (Jarvis, 2005). Consequently, women have tended to become demonized as the culprits of school run congestion precisely because work is gendered and because environmental considerations are, in this case, being prioritized over social inequalities.

Just as the school run is seen as "exceptional" traffic getting in the way of commerce- or employment-related traffic, women's bodies are also seen as "exceptional" to the norm. One of the key arguments for involving women more in environmental decision making at all levels is that their various experiences put them in different and distinctive relationships with the environment. It is likely, from women's own reports, that the experience of carrying and delivering a child contributes to this distinctiveness. As well as hormonal changes, women also become aware of their and their child's vulnerability, particularly when faced with toxic pollution levels from some foods, chemical emissions, and contaminated water (European Environmental Agency, 2003). Compounding this vulnerability is the lack of control of decision-making processes that consistently fail to represent the most vulnerable people in society, just as has been examined above with respect to women. Another example of the importance of the body in analyzing the relationship between gender and environment is the ways in which women's bodies register environmental pollution in different ways than men's (see Corra, 2003, and Dankleman, 2002). While it could be argued that these last two points are essentialist (in that women's biology is part of that woman–environment relationship), it can also be argued that—along the lines of the "social disability model" used to explain how environments and social attitudes create and sustain disability—it is societies' failure to acknowledge and act on multiple bodily responses to pollution that creates such gendered exposure. If, for example, society were to decide that the norm for safe levels of chemical pollution would be determined by the tolerance of a child or a pregnant woman (who are currently advised to take particular care) rather than seeing these as deviant cases from a fit-male norm, then the responsibility for universal environmental protection would become that of the wider society rather than the more vulnerable themselves or their guardians. (See, for example, Parry, 2004, who reported on the failure of major drug trials to adequately represent all end users.)

The foregoing contextual discussion on how gender inequalities pervade legislation, business, and industry, as well as environmental organizations, not to mention economic well-being and social roles, amply illustrates how such inequalities work against "sustainability." In the following section, I review the ways in which a number of international institutions have begun to make a case that gender inequalities have to be redressed in each of these areas if there is to be a chance for socioenvironmental improvement. This is followed by an analysis of how these internationally agreed-upon principles are, or fail to be, incorporated at the level of the individual state through a case study of municipal waste management in Europe, which illustrates how gender relations have been neglected as a cause of environmental damage and as a possibility for achieving socioenvironmental improvements. The case study suggests that change can be achieved when policymakers are receptive to and generate change (microruptures to the fabric of conventional municipal waste management), but that the majority of policymakers are blind to the microgeographies of gender relations and to the potential these offer for socioenvironmental policy change.

THE INSTITUTIONAL CASE FOR GENDER
SENSITIVITY IN SUSTAINABILITY

The World Bank has identified practical reasons, consistent with its aims and practices, for incorporating gender equality into its programs. "Gender is an issue of development effectiveness, not just a matter of political correctness or kindness to women. Evidence demonstrates that when women and men are relatively equal, economies tend to grow faster, the poor move more quickly out of poverty and the well being of men, women and children is enhanced" (World Bank, 2002). Such evidence includes high rates of loan repayments, women's enhanced control over their own fertility and consequent lower birth rates, and directed spending on food, clothes, and other essentials, separately also identified by Chant (1997) and Wickramasinghe (1997).

At the United Nations Conference on Environment and Development in 1992, the Rio Principles and Agenda 21 both confirmed the need to consider gender relations alongside sustainability issues, and this has set the pattern for subsequent UN agreements, including those emanating from the World Summit on Sustainable Development in 2002. Although these sentiments are formally agreed by signatories to the various agreements, the extent to which these are incorporated into national policies may be questioned. Given the unique international legislative capability of the European Union, its global policy commitment to environmental sustainability, and its generally progressive record on human rights and gender equality, it is instructive to follow gender mainstreaming through EU decision-making procedures to examine the impact it is likely to have.

Subsequent to the United Nations Conference on Women in 1995, the European Commission adopted a "Communication on Mainstreaming" in February 1996[1] and agreed that gender impact assessment (GIA) should be a core measure of gender mainstreaming in February 1997. This was formalized in the Treaty of Amsterdam, which promotes positive action, albeit without quotas, in favor of disadvantaged groups in order to achieve equality. The European Commission defines gender mainstreaming as the mechanism whereby

all general policies and measures [be mobilized] specifically for the purpose of achieving equality by actively and openly taking into account at
the planning stage their possible effects on the respective situation of
men and women. This means systematically examining measures and
policies and taking into account such possible effects when defining and
implementing them. (Commission of the European Community, 1996)

The commission recognized that unequal treatment and incentive
measures may be required to secure de facto equality (which does not in-
exorably flow from de jure equality). Structurally gendered differences
affect the development and operation of all policies and the commission
has published guidance on monitoring and impact assessment to enable
policymakers to aim for gender sensitivity and to eliminate unintended
effects. In January 2003, the Council of Europe also agreed a program of
action to promote equal opportunities and to develop gender mainstreaming (Council of Europe, 2003).

The empirical research that informs the rest of this chapter was
commissioned by the European Union's Directorate General for Environment and was designed to pilot gender mainstreaming in one environmental field with a view toward establishing, and making recommendations for, good practice. The research was conducted in the United
Kingdom, Ireland, and Portugal, where there was evidence that these EU
directives had informed policy on gender mainstreaming at the central
government level. Data were collected through emailed questionnaires to
waste management authorities in the three countries (all authorities in
Ireland and Portugal and a sample in the United Kingdom) and through
in-depth analysis of four case studies (interviews with key officers and
councilors, focus groups with people who participated in the local consultations on the waste management plans, and analysis of documentation). Each of the case studies had been chosen on the basis of previous
evidence of good practice in waste management, as the European Union
had specified that one of the purposes of the research was to generate ex-
emplars of gender sensitive waste management practice (further details
of the research can be found in Buckingham, Reeves, & Batchelor, 2005).

Each of the three countries selected for the research had a stated
commitment to equal opportunities and to gender mainstreaming. The
U.K. government had published "Policy Appraisal for Equal Treatment,
Guidelines for Government Departments" in November 1998, and this
was supported by "Gender Impact Assessment: A Framework for Gen-

der Mainstreaming" (Women and Equality Unit, 1998). In Ireland, the Equal Status Act was designed to "promote equality and prohibit types of discrimination, harassment and related behaviour in connection with the provision of services, property and other opportunities" across a range of groups (Department of Justice Equality and Law Reform, 2000: 5). Since October 2001, the Draft National Plan for Women has required gender impact assessment across all policy measures.

Although in Portugal equal opportunities issues have been legislated since 1976, when the Principle of Equality was articulated in the Portuguese Republic's Constitution, the recognition of the need for the integration of a gender perspective on the policy agenda is more recent. Nationally, the adoption of the II National Plan to Equal Rights 2003–2006 offered the possibility of putting this integration into practice in public administration. The Commission for Equality and Women Rights (CIDM) and Commission for Equality in Work and Employment (CITE) are expected to play important roles in implementing and monitoring gender mainstreaming.

Finding evidence for the transmission of these national commitments to equal opportunities and gender mainstreaming, however, was difficult. In the United Kingdom, there was no evidence that government departments (most notably the Department of Environment, Food and Rural Affairs) or agencies (specifically the Environment Agency) were aware of—let alone implementing—good gender mainstreaming practice. Likewise, there was no clear channel through which good practice could be transmitted between central and local government. When each level of government in the three countries was examined for its waste management strategies, only Ireland was found to take gender into account at the national level. At the local level it was found that local authorities' expertise in equal opportunities was not generally utilized in the drawing up of waste management policy.[2] This was as true of substantive issues (as in drawing on expertise in other areas such as transport) as in applying experience of other disadvantaged groups (as in racial/ethnic minorities). Waste management provides a useful vehicle for exploring links between gender inequality, the environment, and sustainability theory. It involves a broad range of activities, both directly and indirectly, that touch many parts of our lives and in which gendered roles are highly significant. Table 3.4 suggests some of the roles and activities and ways in which they are interlinked.

The best example of gender mainstreaming on the ground brought together good practice in employment (where the structure of the waste

TABLE 3.4. Links Between Gendered Roles/Professions and Waste Management Activity

Waste management activity	(Gendered) professional activity	Related (gendered) role
Waste reduction	Female (mainly educational)	Shopping/consumption choices: for example, reusable nappies, lightly packaged goods
Materials reuse	Female (mainly educational; relies on community activity)	Relies on community/ charities activity; reuse within home
Recycling	Female	Women found most likely to recycle
Disposal: land fill and incineration	Male (engineering- and technology-based)	Concerns with environmental health (mostly expressed by mothers)

management team defied the general pattern of male senior management, in which promotion tends to be predicated on a background of waste disposal operations), public participation (where a distinct effort had been made to involve women and men from a range of backgrounds in the process) and good waste management practice (particularly a high emphasis on recycling). These achievements had been generated at the local level not as a result of top-down directives, although national strategies and guidance had been used to facilitate this. Through the interviews, it was clear that one of the critical contributors to this outcome was the appointment of a woman as head of recycling, who drew on her own experience of childcare and domestic responsibilities to instigate and support a "real nappy" campaign as one way in which waste to landfill could be reduced. This example demonstrates how one small change in an appointment can trigger a series of other changes that incrementally may produce gender-sensitive environmental improvements. Table 3.5 demonstrates how these areas of good practice were achieved.

The EU gender mainstreaming research concluded that in what was evidently a highly masculine policy area in all three case study countries, officers, elected representatives, and the public were generally unaware of the likely gendered impact of waste management and often felt uncomfortable with what they thought might be seen as the favoring of a particular group. It was clear that waste management

TABLE 3.5. Good Practice Example of Gender Mainstreaming Waste Management

Activity	Example	Outcomes
Employment practice	Critical mass of women employees, including woman financial manager and head of waste minimization	Authority working through LGA Equalities Standard and working with own Equalities Officer to improve service delivery
Public participation	Consultancy hired to sample population in a stratified way to ensure a representative sample of women invited to public participation and in consultation.	Meetings held at appropriate times/places for a range of people; high response rate to questionnaires
Policy development	Recycling and reuse initiatives, such as real nappies and green cones for composting (low-income families paid to use cloth nappies—reduced Waste Management Authorities' waste disposal costs).	Recycling up to 23% (at a time when the U.K. average was 15%)

planning within the EU, in particular at the local authority level, impacts upon the local community differently according to gender and that gender-differentiated impact was not consistently being taken into account during the stages of designing and implementing municipal waste plans.

Ironically, in most cases, household waste management strategies, which form a critical if generally invisible component of municipal waste management, had not been a consideration of policymakers, who tended to focus on the end product of waste (such as landfill or incineration) rather than its generation. This inattention to the microgeographies of waste generation has implications for the generation of waste that, the research team suggested, could be reduced if waste managers considered the dynamics of waste disposal at the household scale.

Current frameworks for waste management planning design and implementation within the EU are not sufficiently suited to take into account their effects on the respective situation of men versus women, let alone to mitigate the effects. Where equal opportunities expertise had been involved in executive waste management decision making, this appeared to correlate with greater gender sensitivity. Paradoxically, often good practice of sensitivity in other policy areas (both substantively, as in women and public transport, and via process, as in race awareness), failed to be applied to gender mainstreaming in waste management. One conclusion of this research, unsurprisingly, was the recommendation for much more systematic consideration of gender monitoring and transfer of good practice across policy areas, as it is only when practice and experience are transferred that the potential of microruptures can really be materialized.

Linked to this, a number of obstacles to gender-sensitive practice were identified. While all of the intensive case studies had examples of good practice from which other waste management authorities could learn, they were also critically reflective of practices, assumptions, and resource constraints that limited their ability to be more gender-sensitive. One of the main obstacles that was found, particularly with respect to the questionnaire survey of waste management authorities in general and the Portuguese case study in particular, was lack of awareness that gender constitutes sufficient inequality or difference to warrant specific consideration when formulating policy. Arguably, one of the reasons for this is the burial of these inequalities in the microgeographies of the home, where household tasks such as shopping, cleaning, and recycling are still profoundly gendered. In several cases, the research team were asked for help in providing information to enable gender auditing to take place, which suggests that there is considerable scope for awareness raising. Another finding was that the most successful gender-sensitive initiatives emerged when the waste management authority had developed expertise in gender awareness, so that initiatives emerged out of context-sensitive policy discussions rather than being adopted "off-the-shelf" strategies in response to government policy. In particular, there seemed to be a link between the more gender-sensitive waste management strategies, gender balance in waste management appointments, and more effective public participation, which confirms points made earlier in this chapter that there needs to be greater gender equality at senior levels to permeate this equality throughout both the organisation and its policies. In particular, obstacles to gender-sensitive practice were identified and are shown in Table 3.6.

The reported research has suggested that gender mainstreaming has to be thoroughly embedded in an understanding of gender inequalities and differences if it is to be effective. There also has to be a recognition that gender mainstreaming is a continuous process in which policies and practices must be evaluated and the results of these evaluations must in-

TABLE 3.6. Obstacles to Gender-Sensitive Practice in Municipal Waste Management

1. Absorption with other nongendered equal-opportunities issues. Some localities with significant ethnic minority populations have focused all their efforts on raising the participation levels of ethnic minority groups without recognizing that these groups have different gender dynamics.

2. The attitude that "everyone should be treated the same," which may be the product of fear that some residents may accuse councilors of favoritism. This attitude fails to come to grips with ingrained institutional inequalities.

3. The tendency to cater to an "ideal customer" who, in the past, has tended to be male, middle-class, and white (e.g., civic amenity sites have presumed that users are fit, strong, able-bodied, and tall, based on the physical provision of disposal units—likewise the size and weight of waste/recycling bins/boxes). There needs to be a shift to considering how everyone, regardless of differences, needs to have equal access to services.

4. Inadequacy of the public participation process, which may range from its complete absence to not doing enough preparatory work to ensure representation of different groups of women (and men). For example, only one white male adult surveyed had provided childcare facilities to enable parents with no other childcare opportunity to attend a public meeting.

5. Failure to understand that gender differences and inequalities may have an impact on service delivery. This was most likely to be stated by WMAs in Portugal and by elected representatives in the United Kingdom and Ireland, but it could also be identified in officers and members of the public engaged in public participation exercises. It was, nevertheless, interesting to uncover, through interviews and focus groups, how people could perceive differences and inequality once they were prompted to think the matter through.

form future data collection, gender awareness training, and policy development, including the nature of public participation.

CONCLUSIONS

That one of the four case study authorities in the European gender mainstreaming of waste management research had, as a result of local initiatives, managed to introduce measures to reduce environmental impact by being more inclusive of women suggests the rich possibility of linking gender sensitivity with environmental sustainability. If enough of these possibilities can be generated, then we might be able to conceive of each of these as a "microrupture" in the general fabric of a sustainability predicated on sustaining "business as usual." Such a small change in staffing can, however, disrupt structures elsewhere. For women to fairly and effectively perform in the public sphere of the workplace, change is also, and concurrently, needed at the most personal level of domestic arrangements. As long as women are expected to undertake the majority of domestic tasks, this challenges the ability to deliver environmental improvements in a number of ways. For example, the time intensity of combining paid and unpaid work means that those needing to do so are likely to use the most convenient rather than the most environmentally benign strategies. While women are expected to undertake domestic tasks, it is they who will be expected to take responsibility for greening the household, leaving men (and boys) relatively unaffected to continue their environmentally destructive lifestyles. Environmental improvements need 100% of the population's commitment, not 50%. Greater sustainability therefore requires changes in the household as well as the workplace: thus the private and public, the workplace and the domestic space, are imbricated.

"Contraction and convergence"—the concept increasingly being taken up by international agencies—refers to the need to reduce consumption among wealthy states to enable poorer states to raise their standard of living. A similar commitment is needed at the smallest scale so that gender equality can be harnessed to reduce negative environmental impacts on the whole population, not just those who have the power and wealth (as currently obtains) or the visibility (the risk of the current environmental movement) to affect policy.

What the European case studies, dwelt on at some length here, illustrate is that *any* change is difficult to achieve. The research indicated ingrained attitudes toward maintaining the status quo, such that any "positive discrimination" to try to redress centuries of disadvantage was seen as politically untenable, and an inability to understand how environmental issues and policies are structured by gender—indeed a failure to understand gender inequality at all—ensured that gender mainstreaming would be a difficult process to initiate, let alone achieve. An interview with a Portuguese environmental organization captured the incredulity that environmental issues might be gendered when the (male) director responded to a question concerning staff involvement in equal opportunities in waste management as follows:

> ". . . equal opportunities team? Ah, that does not matter! Now, seriously, not at all. That's still extraterrestrial here for our kind. . . . There is no approach to the waste regarding gender. Not only regarding waste . . ."

Clearly, the equal opportunities commitment of Portugal's national government had not been effectively transmitted here. When invited to think about the relevance of the gender dimension, this respondent replied:

> "No, I think that's completely new, and, in the end, taking into consideration our lack of time and resources, and the priorities that are in the line, I sincerely think (and this is a personal opinion, not [name of organisation]) that this is not a priority issue regarding waste. . . . I don't even know what to say, to be quite honest!"

This illustrates the problem referred to earlier in this chapter, that of social and environmental inequalities being regarded in isolation from each other. Environmental NGOs dominated by men in decision-making positions can either wittingly or unwittingly use the environmental dynamic to maintain a silence on the social (in this case gender) dynamic.

Even when good gender-sensitive practice existed in isolated pockets of institutions, such as in human resources, there was little evidence that anyone had thought to transmit this practice across to other sectors. Transferring good gender mainstreaming practice across scales, too, was seen as problematic and lacking structures through which this could be attempted. Such institutional barriers ensure that any achievement toward greater gender sensitivity in environmental policy and practice remains isolated. Over time, this might achieve a degree of incrementalism, but this is unlikely to have a significant impact on how environmental policy may be sufficiently changed to tackle fundamental and structurally embedded problems unless there are mechanisms and people who can make the necessary links to jump policy silos.

My conclusion from this and other research discussed at the beginning of this chapter is that legislative and policy changes alone are insufficient to the task of forging an approach to intertwined environmental and social inequalities that will enable these to be redressed. It may be sufficient to the task of "sustainability"—that is, making small-scale incremental changes within the prevailing status quo that will deliver precisely the insufficient reductions in greenhouse gases currently obtaining or that will encourage public acceptance of nuclear power as the preferred alternative to fossil fuels. For the significant change that is needed, many more durable "microruptures" must tear at the fabric of "sustainability" and challenge us to look at a future in which the relationship between gender (and other structures of disadvantage) and environment is constituted differently. Before we get to "sustainability" as a normative aspiration, we need significantly more social–environmental change to ensure that what we are sustaining is of value to the whole community. While, following Lipietz, we should not lose faith in the ability of small changes to generate larger-scale change, "microruptures," such as the few reported in the case study here, must occur at all scales in order to overcome the lack of transmission paths between scales. Nothing short of a *reconstruction* of gender relationships, and consequently of society–environment relations, is required to achieve a condition that is worth sustaining.

NOTES

1. "Communication on Mainstreaming" final of February 21, 1996, on "Incorporating equal opportunities for women and men into all community policies and activities."
2. Waste management authorities' boundaries are not always congruent with those of local authorities.

REFERENCES

Barker, J. (2006). *Are we there yet?: Exploring aspects of automobility in children's lives*. Unpublished PhD thesis, Brunel University, West London, UK.

Bhattar, G. (2001). Of geese and ganders: Mainstreaming gender in the context of sustainable human development. *Journal of Gender Studies, 10*(1), 17–32.

Bradshaw, J., Finch, N., Kemp, P. A., Mayhew, E., & Williams, J. (2003). *Gender and poverty in Britain*. Manchester, UK: Equal Opportunities Commission.

Buckingham-Hatfield, S. (2000). *Gender and the environment*. London: Routledge.

Buckingham, S., Budd, J., Lynn, H., Murphy, D., & Sutton, L. (2005). *Why women and the environment?* London: Women's Environmental Network.

Buckingham, S., & Lievesley, G. (Eds.). (2006). *In the hands of women: Paradigms of citizenship*. Manchester, UK: Manchester University Press.

Buckingham, S., Reeves, D., & Batchelor, A. (2005). Wasting women: The environmental justice of including women in municipal waste management. *Local Environment, 10*(4), 427–444.

Butler, J. (1990). *Gender trouble: Feminism and the subversion of identity*. London: Routledge.

Carter, N. (2001). *The politics of the environment: Ideas, activism, policy*. Cambridge, UK: Cambridge University Press.

Chant, S. (1997). *Women-headed households: Diversity and dynamics in the developing world*. London: Macmillan.

Climate for Change. (2002). *Gender equality and climate policy.* Accessed October 19, 2006, from *http://www.climateforchange.net/22.html.*

Commission of the European Community. (1996). *Gender mainstreaming.* Brussels: CEC.

Commission of the European Community. (1997). *Treaty of Amsterdam.* Accessed February 2, 2004, at *http://www.europa.eu.int/scadplus/leg/en/chc/c10101.htm.*

Council of Europe. (2003, January 22–23). Report of the 5th Ministerial Conference on Equality between Women and Men. *Skopje,* MEG-5 Strasbourg; Council of Europe Directorate General of Human Rights.

Commonwealth Secretariat. (1998). *Gender mainstreaming: Commonwealth strategies on politics, macroeconomics and human rights.* London: Commonwealth Secretariat.

Corra, L. (2003, Autumn). Speech reported in *WEN News.* London: Women's Environmental Network.

Dankleman, I. (2002, March 14–17). Poverty eradication as a challenge for sustainable development. In *Women in Europe for a common future: Why women are essential for sustainable development* (pp. 68–71). Proceedings of the European Women's Conference for a Sustainable Future, Celakovice (Prague), Austria.

Deloitte & Touche LLP. (2006). *Executive directors decline in number as FTSE boards shrink to meet Higgs code.* Accessed October 15, 2006, at *http://www.deloitte. com/ dtt/press_release/0,1014,sid%253D2992%2526cid% 253D130907,00. html.*

Department for Environment, Food and Rural Affairs. (2002). *Survey of public attitudes to quality of life and to the environment—2001.* London: Author.

Department for Environment, Food and Rural Affairs. (2005). *The UK government sustainability strategy,* Command Numbers 6467. London: Author.

Department for Transport. (2005). *Focus on personal travel.* London: The Stationery Office.

Department of Justice, Equality and Law Reform, Ireland. (2000). *Equal Status Act.* Dublin: Author.

Dowling, R. (2000). Cultures of mothering and car use in suburban Sydney: A preliminary investigation. *Geoforum, 31,* 345–353.

Equal Opportunities Commission. (2006). Accessed August 21, 2008, at *http://www.eoc. org.uk/Default.aspx?page=17922.*

European BoardWomen Monitor. (2004, June). *Another north–south divide: Women on boards* (press release on the launch of the European Professional Women's Network, citing a survey of the top 200 European industrial and service companies, based on revenue, published by Les Echos and Reuters in *Le Grand Atlas des Entreprises,* Paris, June, 14, 2004.

European Commission, Employment and Social Affairs. (2004). *Positions held by women and men in European Union institutions.* Brussels, Belgium: Author.

European Environment Agency. (2003). *Environmental Assessment Report.* Copenhagen, Denmark: Author.

Friends of the Earth Scotland. (2000). *The campaign for environmental justice.* Edinburgh, UK: Author.

Greed, C. (1994). *Women and planning: Creating gendered realities.* London: Routledge.

The Guardian. (2006, October 2). Accessed October 15, 2006, at *business.guardian.co.uk/story/0,,1885252,00.html.*

Her Majesty's Government. (1999). *Social trends.* Accessed August *, 2006, at *http:// www. statistics.gov.uk/StatBase/Expodata/Spreadsheets/D3700.xls.*

Hunt, K. (2006). "Embedded Feminism" and that War on Terror. In K. Hunt & K. Rygiel (Eds.), *Gendering the War on Terror, war stories and camouflaged politics*. Aldershot: Ashgate.

Jarvis, H. (2005). *Work/life city limits: Comparative household perspectives*. Basingstoke, UK: Palgrave.

Kingsmill, D. (2001). *Kingsmill report on women's employment and pay*. London: Department of Trade and Industry, UK.

Law, R. (1999). Beyond 'women and transport': Towards new geographies of gender and daily mobility. *Progress in Human Geography, 23*(4), 567–588.

Lievesley, G. (2004). *The Cuban revolution: Past, present and future perspectives*. Basingstoke, UK: Palgrave.

Lievesley, G. (2006a). Women and the experience of citizenship. In S. Buckingham & G. Lievesley, *In the hands of women: Paradigms of citizenship*. Manchester, UK: Manchester University Press.

Lievesley, G. (2006b). Identity, gender and citizenship: Women in Latin America and Cuba. In S. Buckingham & G. Lievesley, *In the Hands of Women: Paradigms of citizenship*. Manchester, UK: Manchester University Press.

Lipietz, A. (1995). *Green hopes. The future of political ecology*. Cambridge, UK: Policy Press.

Lipietz, A. (2000). Political ecology and the future of Marxism. *Capitalism, Nature, Socialism, 11*(1), electronic version, unpaged.

Lister, R. (1997). *Citizenship: Feminist perspectives*. Basingstoke, UK: Palgrave.

MacGregor, S. (2006). No sustainability without justice: A feminist critique of environmental citizenship. In A. Dobson & D. Bell (Eds.), *Environmental citizenship*. Cambridge, MA: MIT Press.

Marston, S. (2000). The social construction of scale. *Progress in Human Geography, 24*(2), 219–242.

Mehdid, M. (1996). En-gendering the nation-state: Women, patriarchy and politics in Algeria. In S. M. Rai & G. Lievesley, *Women and the state: International perspectives*. London: Taylor & Francis.

Mellor, M. (1992). *Breaking the boundaries: Towards a feminist green socialism*. London: Virago.

Merchant, C. (1996). *Earthcare, women and the environment*. London: Routledge.

Mies, M., & Shiva, V. (1993). *Ecofeminism*. London: Zed Books.

National Anti-Hunger Organizations. (2004). *A Blueprint to End Hunger* (a working document based on the principles of the Millennium Declaration to End Hunger in America, issued by NAHO).

National Commission on Pay Equity. (2005) Accessed April 18, 2005, at *http://www.pay-equity.org/index.html*.

Nightingale, A. (2006). The nature of gender: Work, gender, and environment. *Environment and Planning D: Society and Space, 24*, 165–185.

Parry, V. (2004, November 3). Let's look at those tests again. *The Guardian*.

Rootes, C. (1999). Environmental movements: From the local to the global. *Environmental Politics, 8*(1), 1–12.

Seager, J. (1993). *Earth follies: Coming to feminist terms with the global environmental crisis* London: Routledge.

Staeheli, L. (1994). Empowering political struggle: Spaces and scales of resistance. *Political Geography, 13*, 387–391.

Swyngedouw, E., & Heynan, N. C. (2003). Urban political ecology: Justice and the politics of scale. *Antipode, 35*(5), 898–918.

United Nations Millennium Project. (2005). *Taking action: Achieving gender equality and empowering women.* Task Force on Education and Gender Equality. London: Earthscan.

Visvanathan, S. (1991). Mrs. Brundtland's disenchanted cosmos. *Alternatives, 16*(3), 377–384.

Wickramasinge, A. (1997). *Land and forestry: Women's local resource based occupations for survival in South Asia.* Peradinaya, Sri Lanka: Corrensa.

Women's Environmental Network. (1993). *Women's Environmental Network Campaigns: 5 Years of WEN.* London: Author.

Women and Equality Unit. (1998). *Gender impact assessment: A framework for gender mainstreaming.* Available at *http://www.womenandequalityunit.gov.uk.*

Women's Resource Centre. (2006). *Why women? The women's voluntary and community sector: Changing lives, changing communities, changing society.* London: Author.

World Bank. (2002). *Gender mainstreaming strategy paper.* Accessed October 23, 2006, at *http://www.worldbank.org/gender/overview/ssp/home/htm.*

CHAPTER 4

Containing the Contradictions of Rapid Development?

New Economy Spaces
and Sustainable Urban Development

DAVID GIBBS
ROB KRUEGER

In recent years we have seen the emergence of two phenomena that indicate major shifts are under way in the organization of economies and societies: the development of a *new economy* and incipient efforts to promote *sustainable development*. The development of a *new economy* composed of high-technology sectors and knowledge-based "production," such as information and communication technologies and biotechnology and the so-called FIRE sector, which includes finance, insurance, and real estate, is both transforming production and consumption norms as well as altering relations among business organizations, individuals, and institutions (Storper, 1997; Scott, 1987; Porter, 1990; Ley, 1996; Thrift & Olds, 1996; Scott, 2000; Nevarez, 2003; Gleeson & Low, 2000). As with previous rounds of economic development, this new economy is concentrated into specific locales—in this case, into a number of city-regions. Indeed, a parallel set of arguments has emerged suggesting that such city-regions have become the locus of not only the new economy but also global economic growth (Herrshel &

Newman, 2002; Scott, 2001). The success of these "new economy spaces" has meant that local policymakers have sought to replicate similar conditions in less successful locales in the continuing elusive search for economic development (Krueger & Buckingham, 2005; Kong, 2000). To this end, numerous academic and popular works have investigated and/or advocated the merits of developing "clusters" (Porter, 2000), "innovative milieux" (Scott, 2000; Nevarez, 2003), "creative cities" (Kong, 2000; Landry, 2000), and the "creative class" (Florida, 2002) as elements of local economic development strategies.

In parallel with the growing policy and conceptual importance of the new economy, the concept of sustainable development also seems to be a key part of the policy mix for urban and local governments. Sustainable development is a discourse that seeks to offer a somewhat different view of future economic and social organization from that of new economy analysts (World Commission on Environment and Development, 1987; O'Riordan, 1999). Here, notions of combining economic, environmental, and social development in a holistic manner have emerged along a spectrum of approaches from light to dark green, ranging from business as usual with a "green tinge" to deep ecology approaches (Gibbs, 2002; Luke, 1996). Similar to research on the new economy, there has been an outpouring of work on sustainability at the local and urban scale (see Gibbs, 2002; Gibbs & Krueger, 2004), with policy prescriptions for "ecological cities" (Platt, 2004), "compact cities" (Breheny, 1995), "green urbanism" (Beatley, 2000), and measuring "ecological footprints" (Wackernagel & Rees, 1996) and "industrial ecology" (Gibbs, Deutz, & Proctor, 2005). There is considerable evidence that local sustainability initiatives are growing in number—the International Council for Local Environmental Initiatives (ICLEI; 2002), for example, reports that more than 6,000 communities worldwide have adopted sustainability planning practices.

At first sight, these two developments may seem antithetical to each other. Capitalist growth is frequently associated with environmental degradation, thereby creating "both a material crisis of production and a legitimization crisis for capital" (Angel, 2000: 611). Indeed, O'Connor's (1998) second contradiction of capitalism proposes that capitalist development will tend to degrade the ecological conditions it depends upon. In this view, a tendency toward ecological crisis is just as endemic to capitalism as a falling rate of profit or overaccumulation. At first sight, then, we might expect the new economy to differ little in its impact upon the environment from older forms of economic development. Indeed, some research would support this conclusion—for example, Pellow and Park's (2002) research on the archetypal new economy of Silicon Valley indicates that its high-tech sectors may be a significant local pollution source. However, the reality may be more complicated than this suggests. The limited, and often anecdotal, evidence to the contrary that exists suggests that on some indicators the top performers in the new economy are also leading exponents of sustainable development (see Table 4.1). For us, this raises a number of key questions. First, do material conditions in new-economy spaces support both a model of global competitiveness *and* the principles and practices of sustainable development? Second, is a concern for sustainability integral to the development of these new-economy spaces? Third, how is sustainability constructed in terms of policy goals in these locales? Finally, what institutional forms have evolved in new-economy spaces to address these issues? In this chapter we provide an exploratory examination of these questions by focusing on two empirical case studies: Austin, Texas, and Boston, two of the leading locations for the new economy in the United States (see Table 4.1). Our purpose then in this chapter is to consider some conceptual

TABLE 4.1. Top 10 New Economy Spaces and Sustainable Development Cities in the United States

Top 10 on the New Economy Index[a]	Top 10 sustainable cities on Sustainlane Index[b]	Top 10 green cities in *thegreenguide.com* index[c]
1. San Francisco, CA	1. Portland, OR	1. Eugene, OR
2. **Austin, TX**	2. San Francisco, CA	2. **Austin, TX**
3. Seattle, WA	3. Seattle, WA	3. Portland, OR
4. Raleigh–Durham, NC	4. Chicago	4. St. Paul, MN
5. San Diego, CA	5. Oakland, CA	5. Santa Rosa, CA
6. Washington, DC	6. New York, NY	6. Oakland, CA
7. Denver, CO	7. **Boston, MA**	7. Berkeley, CA
8. **Boston, MA**	8. Philadelphia, PA	8. Honolulu, HI
9. Salt Lake City, UT	9. Denver, CO	9. Huntsville, AL
10. Minneapolis, MN	10. Minneapolis, MN	10. Denver, CO

[a]This column ranks metropolitan areas on the basis of five categories of indicators: knowledge jobs, globalization, economic dynamism and competition, transformation to a digital economy, and technological innovative capacity. Rankings are for 2001. For details, see *http://www.neweconomyindex.org*.

[b]Rankings for 2006. Cities ranked on a range of quality of life and sustainability indicators. For details, see *http://www.sustainlane.com*.

[c]Rankings for 2006 and based upon a combination of rankings for environmental policy, environmental perspective, green design, green space, and public health; air quality; electricity use and production; recycling; water quality; socioeconomic factors; and transportation. For details, see *http://www.the greenguide.com*.

linkages between sustainability and the new economy in order to shed some light on how sustainability might be congruent (or not) with neoliberal capitalist forms.

The structure of the chapter is as follows. Drawing on existing theory and conceptual work, in the next section we develop the argument that a closer examination of the interrelationship between the two phenomena raises a set of questions that need to be addressed. We then outline the key environmental and quality-of-life issues that are of concern within new-economy spaces, before turning to our empirical investigations of Austin and Boston. We draw upon secondary sources and published literature for our analysis in this section, as well as interviews conducted with key stakeholders in both areas.[1] Following a section that relates these empirical findings to our theoretical arguments, we come to some conclusions on the potential for a future research agenda.

THEORIZING SUSTAINABILITY: FROM IDEALS TO EQUITY AND BEYOND

Sustainable development is often defined as a process that integrates the three domains of environment, economy, and society, with sustainable development seen as the intersection between these, as depicted in classic Venn diagram format (O'Riordan, 1999). Since the Brundtland report was published in 1987, sustainable development has increasingly become an important discourse in policy debates at all spatial scales. The Brundtland mantra of not leaving people in the future worse off has found its way into academic and policy work worldwide. Despite its widespread adoption, until a couple of years ago many applications were normative in nature and focused primarily on the environment. Recently, these definitions of sustainable development produced by both scholars and practitioners have become increasingly sophisticated (see Hempel, 1999; Agyeman, Bullard, & Evans, 2003; Buckingham & Theobald, 2003; HM Government, 2005), especially in terms of broadening the scope of who should be included in sustainable development. These definitions, while maintaining the spirit of Brundtland, are much more pragmatic. Indeed they seek to address specific urban problems associated with the neglected negative externalities of modern city making. Here, discussions of environmental limits per se are substituted for issues of social and environmental justice. This work rises to Haughton's

(1999) challenge of acknowledging the interdependency of social justice, economic well-being, and environmental stewardship. He argues that "the social dimension is a crucial one since the unjust society is unlikely to be sustainable in environmental or economic terms in the long run" (1994: 64). This represents a significant shift away from the kinds of trade-offs between economy and environment in past policy definitions (Gibbs, 2002). In this new work on urban sustainability the conversation implicitly acknowledges the human–environment interaction in very real and specific ways. Rather than concerning themselves with ecological footprints and sustainable cities writ large, increasingly authors have engaged with sustainability as a justice issue related to problems involving specific social groups, such as women and disproportionately affected groups (see Agyeman, 2005; Buckingham & Lievesley, 2006). Despite the merits of these conceptual and (limited) practical interventions, the current state of thinking suggests that we remain in an impasse between recognizing the need for policy and the ability to deliver it (Gibbs, 2002; Krueger & Agyeman, 2005).

While the sustainability literature has certainly evolved in the past few years to include exploring issues of equity beyond a focus upon environmental concerns, there is still more work to be done. In the rest of the chapter we explore a set of theoretical concerns relating to social change that we believe commentators on sustainability need to engage with. In the absence of concepts such as materiality, power relations and hegemony, multiple constructions of sustainability, and changes to institutional form and function under different political economic conditions, commentaries on sustainability are reduced to something less than they could be. For without exploring the roots of injustice or the realms of the possible under current political economic conditions, how can we expect to make more progressive changes to the system? The chapter now turns to engage with the four questions set out in our introductory section.

CONVERGENT CULTURES?: GOVERNING ECONOMIC DEVELOPMENT IN THE NEW ECONOMY

The new economy is widely recognized as forming the main driver of global economic growth in the developed world (see, among others, Kelly, 1998; Watson, 2001; Daniels, Beaverstock, Bradshaw, & Leyshon, 2005). According to the Progressive Policy Institute (Atkinson, 1999), the new economy is a "knowledge and idea based economy where the keys to wealth and job creation are the export of ideas, innovation and technology that are embedded in all sectors of the economy." While the new economy is composed spatially of groups of high-tech industries, such as information and communications technology, biotechnology, nanotechnology, and so on, there is also a wider sense that a broader "new economy" is developing that encompasses changes in the way that business organizations operate and relate to other institutions, both internally and externally (Kelly, 1998), which has implications for the local areas that "house" these new-economy firms. Moreover, these new relations could have implications for spatial governance strategies.

New-economy spaces mark a new "sphere of convergence" between economic activity and culture. In these spaces capitalism is "moving into a phase in which cultural forms and meanings of its outputs become critical if not dominating elements" (Scott, 1997: 323). Economy and culture have always been intrinsically interwoven, but as Scott (1997) and others point out, new economies involve a much closer relationship between the firm and place, such that new forms of economic activity reflect a dialectic between local culture and capital. We are already familiar with notions of firm embeddedness in local places (Storper, 1997), which takes a variety of forms including internal organizational changes that involve breaking down departmental silos and altered business-to-business relationships, but the connectedness between firms and place and between capital and culture are moving into new realms. Thus, new-economy firms have different relationships with their host locations. In contrast to their Fordist predecessors, which were interested in public subsidies, tax breaks, inexpensive real estate, few regulations, and low wages, it is argued that new-economy firms view location in terms of linkages to external economies (Saxenian, 1994; Scott, 1997), specialized business services (Sassen, 1991), and destination spaces (Judd, 1999; Zukin, 1995) rather than focusing solely on more tangible considerations such as operating costs and rents (Nevarez, 2003). New-economy spaces, theoretically at least, thus have a synergistic relationship with their firms.

It is for this reason that we can begin to address the question of why new economy spaces may also take sustainable development seriously as an integral part of their development. Thus, the extended relationship between firms and the cultural qualities of place is particularly evident in relation to quality-of-life issues and environmental assets in new-economy spaces. Both quality-of-life attributes and environmental assets are deemed essential in the new economy to maintain competitiveness vis-à-vis other locations (Bluestone, 2006). In particular, attracting and retaining highly qualified and highly paid key workers is closely linked with quality-of-life issues, such as attractive landscapes and opportunities for leisure (Nevarez, 2003; Florida, 2002; Walker, 2003; Gottlieb, 1995; Herzog & Schlottman, 1991). Elite workers are especially sensitive to local quality-of-life issues insofar as their particular training and skills often give them greater spatial choice and mobility than their less well paid, nevertheless colocated, counterparts. Indeed, the longevity of new-economy spaces may ultimately be determined by the local dependence of elite workers rather than that of the firms for which they work (Cox & Mair, 1988; Nevarez, 2003). Quality of life is thus important in competitiveness terms because it both attracts and helps to retain both new-economy workers and firms within areas (Saxenian, 1994; Gottlieb, 1995; Atkinson, 2002).

NEW-ECONOMY PROBLEMS, NEW-ECONOMY GOVERNANCE

However, new-economy spaces do not merely represent an unproblematic, even convivial, set of relationships between firms, the state, and the environment. Here we can begin to address the question of the form that a concern for sustainability takes in new-economy spaces. Many times it is the contradictions inherent to new-economy spaces, in particular those arising from development (both commercial and housing), environment, and quality of life in new-economy spaces that become the crucible of engagement. Quality of life, for example, typically requires maintaining the fragile balance between economic competitiveness, social and environmental amenities, and affordable housing. As O'Connor's second contradiction of capitalism might suggest, economic "success" measured in conventional GDP terms frequently leads to traffic congestion, poor air quality, groundwater pollution, stress on water reserves, and loss of landscape amenities. Thus, as Prytherch (2002: 773) points out in his study of Tucson, "Marketers may construct nature as a 'condition' for the production of growth, but sprawl devours the landscape upon which their sales pitch is premised." Physical and social infrastructures within new-economy spaces are pressured by increasing land scarcity, traffic congestion, house price inflation, and demands on the local tax base (Harvey, 1985; Saxenian, 1994; While, Jonas, & Gibbs, 2004; Gibbs & Krueger, 2004).

In absolute terms, the direct demands for space from high-tech activities are relatively small as compared to the demands from other associated
uses. For example, in Cambridge, United Kingdom, it is estimated that for
every hectare of land required for high-tech business an additional 15 hectares is required for related housing, physical, and social infrastructure
(Segal, Quince, & Wicksteed, 2000). The impacts of growth pressures are
therefore more closely related to the need to reproduce conditions in the
living space, such as maintaining the quality of the local environment, facilitating the supply of labor through housing, transport, schools, and
health provision, and more generally meeting the demands of local residents for adequate services (Cox & Jonas, 1993). Moreover, such growth
is often accompanied by social polarization, labor turnover, and worker
discontent (Allen, Massey, & Cochrane, 1998; Crang & Martin, 1991).
Social tensions arise in the context of many lower-paid support workers in
more mundane jobs. New economy space "success" may price them out of
local housing markets, leading to long-distance commuting, two-income
households, and negative impacts upon both environmental pollution and
family life (Walker, 2003; Luke, 2003).

Institutional responses to these contradictions also represent a key
area of inquiry. What institutional forms and practices evolve to address
these tensions between the consequences of growth and the preservation
of quality of life? The regulations, decisions, and policies affecting sustainable outcomes at the local level are constructed through discursive
practices and struggle over materiosocial structures (Gibbs & Jonas,
2000; Krueger, 2002). For Molotch (1996), one way new-economy firms
have sought closer forms of engagement with state forms is through public–
private partnerships. In this sense, then, such new economic spaces may
be closely bound up with the development of new forms of institutional
and corporate cultures. The key issue here revolves around structures of
governance and institutional forms that arise to address the tensions between growth, social equity, and preservation of local amenities and the
environment. Here, the processes and structures of governance and regulation become critical analytical issues. We draw here upon the work of
Jessop (1990, 1995, 2002), who conceptualizes the state as an "institutional ensemble" and characterizes state power as reflecting the interrelationships between the interests of politicians and state managers and
the promotion of interests by social and economic forces (Jonas, Gibbs,
& While, 2004). Jessop's neo-Gramscian concept of "strategic selectivity" suggests that some actors and institutions have the ability to formu-
late, secure, and implement specific policies while others do not. Spe-

cifically, the terms of reference for what is being "strategically selected" in the first place largely rests with elite groups in society. While such strategic selectivity in most developed states continues to prioritize competitiveness, entrepreneurialism, and a largely neoliberal agenda over sustainability concerns, this may not be tenable or desirable in new economy spaces. As Nevarez (2003: 11) points out, elite groups in new economy spaces have a different relationship to the political process and the economic landscape than their predecessors did—"a company has to work out of a physical setting, which means it has particular needs from a locality and engages in certain relationships and activities in order to do business there."

In new-economy spaces, therefore, a concern for sustainability issues may not be an *obstacle* to capitalist accumulation but rather a *constituent part* of it. The consumption of resources may push elites in new economy spaces to promote social and environmental programs, and, as Nevarez (2003) shows for parts of California, forms of governance in such spaces are driven by different corporate cultural values. As we have argued, in the spaces of the new economy, economic success, quality of life, and a "good environment" are closely intertwined. Local political elites need to address the tensions and policy dilemmas that arise in new economic spaces if they are to secure the continued success of their local areas. One key question here is the extent to which elites in new-economy spaces are shifting away from "strategically selective" approaches that focus on competitiveness and entrepreneurialism and toward new forms of social regulation based on equity and quality-of-life issues. In order to address these issues, we now turn to an examination of two U.S. cities that are at the forefront of high-tech development and with long-standing environmental credentials.

AUSTIN, TEXAS: GREENING THE NEW ECONOMY CITY

The city of Austin has expanded rapidly in recent years, with a population growth of 42.8% between 1994 and 2004, leading to a total population of 1.4 million people in the metropolitan statistical area (MSA). The catalyst for this rapid population growth is its role as a leading location for high-tech industry, including ICT and biotechnology firms, with the Dell Corporation also headquartered in the city (McCann, 2003). This high-tech boom had its roots in the local development strategies of the 1950s, reaching its height during the 1980s and 1990s and the dot.com boom, when Austin was acclaimed as a high-tech capital that also boasted a high quality of life (McCann, 2005). At the same time, the city has gained a reputation for having a progressive approach to environmental issues:

> The City of Austin is among the elite when it comes to setting environmental policy. Over the past fifteen to twenty years the City of Austin has earned hundreds of environmental awards. The following City programs have received national, state, or trade association recognition: Water/Wastewater Department's Dillo Dirt, Keep Austin Beautiful, Water Conservation, Austin Recycles, Energy Conservation, Public Works, Green Builders, and the Propane Program. Other active City-sponsored environmental programs include alternative fuel technology, teleworking, alternative commuting, tree planting, sustainable communities and "smart growth" incentives. These environmental initiatives are not just government-led; Austin is one of the most environmentally active communities in the country: The City of Austin was ranked second "greenest" city in the nation by the World Resources Institute in 2004. There is a powerful environmental coalition of organizations, with active local chapters of the Sierra Club, National Audubon Society, Environmental Defense, and over fifty other environmental organizations in the city alone. (Gunn, 2004: 7–8)

This strong local environmental culture in the city, together with a mix of environmental and other interest groups, both supports environmental preservation and protection and has an important input into local policymaking (Gunn, 2004). From 1996 onward, the city developed a Sustainable Communities Initiative that created plans, conducted evaluations, and educated city staff and the public on ways to make the city more sustainable. The Austin Green Builder Program was the first sustainable residential program developed in the United States, and this has now spread into other areas of construction and development (Tinker, 2003). More recently a 2005 transit-oriented development ordinance aims to encourage maximum mixed-use density development around major transit nodes (Nichols, 2006). One of the primary roles for policymakers and local government leaders in Austin has been to focus on quality of life. In the mid-1980s an independently commissioned quality-of-life survey rated the city as "exceptional" (as compared to San Diego, Atlanta, and Raleigh–Durham) in terms of the quality of primary and secondary schools, the quality of its parks and playgrounds, outdoor recreational opportunities, community cleanliness, and as an affordable place to live (Smilor, Gibson, & Kozmetsky, 1988). These are seen as key assets in attracting the kinds of highly qualified labor that the city's high-tech industries rely upon.

Despite these plaudits, Austin suffers from the same problem as many other high-tech locations, namely, how to continue economic development but at the same time maintain the local environment and quality of life. As Smilor et al. (1988) point out:

> Over the history of the economic development of the Austin area, local government has tended to favour either the "developers" or the "environmentalists." When local government supports economic growth then the development of the technopolis is more likely to increase; that is, company relocation seems to be facilitated and obstacles to development seem to diminish. On the other hand, when local government believes that the quality of life is diminishing, then the development of the technopolis is inhibited; that is, obstacles to development increase (such as high utility rates or slow permit procedures). The issues become quite complex because many developers are often local residents who also want to preserve the community's quality of life.

One of the key battles within the city has come over the location of any future development, with concerns over the impacts upon the Texas Hill country to the west of the city and, in particular, over the impacts upon the Barton Springs portion of the Edwards aquifer. While some developers and firms have wanted more office and manufacturing space and housing developments, these have been opposed by environmentalists—particularly the Save Our Springs Alliance—and some businesses concerned about the loss of habitat for endangered species and the aquifer from which the city obtains much of its drinking water (McCann, 2005). The 1990s was a period when "a coalition of environmentalists . . . organized around issues of environmental justice and pollution in inner city neighborhoods, engaged in conflict with developers and newly arrived corporations over the future of urban development" (McCann, 2003: 165). Local elections at this time produced a Democrat-dominated "green city council" that sought to address four interrelated problems: population growth, the site demands of high-tech firms, social and economic inequality, and the environment (McCann, 2003). Policymakers have attempted to address these issues in various ways. In 1997 Austin developed a Smart Growth Initiative, effectively a set of land-use policies, which was intended to discourage development in the environmentally sensitive areas while promoting growth in the urban core, especially in the downtown area. It was specifically intended to determine how and where development should take place, improve the quality of life, and enhance the city's tax base. It created a Desired Development Zone and a Drinking Water Protection Zone, where development was to be encouraged and discouraged, respectively (City of Austin, 2001). However, Smart Growth in Austin was an incentive-based rather than regulatory response to urban growth, in part influenced by the local context, where clashes with the state government had "led to a series of disputes over development between the traditionally liberal and environmentally conscious city of Austin and the traditionally conservative and growth-oriented state of Texas" (McCann, 2003: 168).

While it proved possible to manage the tensions (or produce a "sustainability fix") between environment and development for a time, this became increasingly problematic following the impact of the dot.com crash that followed the boom of the 1990s. While population grew by 47.7% between 1990 and 2000, it subsequently slowed to 13% between 2000 and 2004 (Greater Austin Chamber of Commerce, 2005). At the same time, unemployment rose, and the city's revenues from both sales and corporate taxes declined significantly. The fragile consensus between environmentalists, developers, and neighborhoods that had been created through Smart Growth fell apart as environmental groups favored a "no growth" approach as opposed to the Smart Growth plan of directed growth, and developers left building projects uncompleted in the downtown area to the annoyance of local activists for whom city center redevelopment was a major factor in combating urban sprawl (interview, Smart Growth coordinator, December 2003). The negotiated consensus over protecting the aquifer has also started to unravel, with debate over environmental protection versus economic development continuing with the controversy over attempts by Advanced Micro Devices to build a corporate campus on the Barton Springs watershed and opposition by the Save Our Springs Alliance (*Austin Chronicle*, *www.austinchronicle.com/issues*, accessed January 27, 2006).

It has also become increasingly apparent that the downside to Austin's growth has been an increase in inequality and a decrease in housing affordability, particularly among the city's African American and Latino populations (McCann, 2005). This inequality has also been the subject of local debate both among activists and the business community. As McCann (2005: 12) points out:

> While late-Twentieth Century Austin gained an image as a high-tech boomtown and an ideal hometown where Creatives could "live the life," the city's politics and policy were dominated by ongoing negotiations between the local, state and various activist groups aimed at mitigating the negative effects of rapid urban growth on fragile landscapes and on low-income people.

Indeed, although they may disagree over the most suitable policy to reconcile economic development and environmental policy, both environmentalists and developers "agree that overall quality of life suffers when the people who inhabit the community are out of work and cannot afford to pay the costs associated with infrastructure development, housing, or factors such as expanded park land or recreational opportunities" (Smilor et al., 1988). As McCann (2005: 14) points out, debates over quality of life and "livability" should not be taken for granted: "Rather than being a self-evident and generally agreed upon 'fix' to institutional and geographical problems of urban development, the regionalist livability agenda has become the context and object of a wide range of urban political struggles."

Within Austin there has been a search for new forms of governance and new institutions that can adequately address both the perceived development needs of the city and yet, at the same time, maintain a high quality of life and address some of the inequality problems. Two specific local institutionalist approaches or intended "fixes" for these problems have come in the form of the Austin Network and Opportunity Austin. The Austin Network arose out of a local conference, the Austin 360.00 Summit, that brought together many of the city's high-tech CEOs in response to the questions of preserving quality of life, improving infrastructure, and an apparent disconnect between the technology community and community issues (Bishop, 2000). Some have argued that this type of development effectively represents a new form of governance or a "network governance mode" (Bishop, 2000) in high-tech areas, where voluntary associations of business leaders, nonprofit groups, and local government come together to address issues that may spill over formal government boundaries, blurring the lines between the private and public sectors. In Austin, part of the strategy involves the use of technology in the form of GetHeard.org, a website that provides a channel of communication between the private high-tech sector, local government officials, and nonprofit organizations. As elsewhere, public–private partnerships have become one way to address governance problems.

Opportunity Austin was a direct institutional response to the loss of over 30,000 jobs after the 1990s boom, combined with a national recession and technology shifts. These changes led the local chamber of commerce to commission a report on the city's economy, to benchmark it against competitors and to develop a job creation strategy. This resulted in the creation of Opportunity Austin, with the aim of creating 72,000 jobs and an increase in payrolls. As a consequence, the Greater Austin Economic Development Corporation was created to oversee the initiative's planning and progress and to seek financial support. Portfolio Austin—A Strategy for Growth was launched in August 2004, with a central component composed of a " 'SWAT team' of regional leaders to respond to critical challenges standing in the way of local businesses' growth and success" (Opportunity Austin, 2005). Despite the economic focus of this institutional and governance response, Opportunity Austin continues to prioritize environmental protection; thus, "We must protect the local quality of life—so citizens can move about quickly, cheaply and conveniently, enjoy clean air and water, and take advantage of the region's abundant green space" (2005: 2). Indeed, environmental issues form a central part of Austin's development plans. Thus, Opportunity Austin's economic development plan includes a strategy to encourage the development of a clean energy sector in the city through a newly established Clean Energy Development Council (CEDC), funded and endorsed by the city council and the chamber of commerce. This is promoted as a win–win between economic development and the environment, although skeptics are said to be "watching to see if the CEDC is indeed an economically and environmentally sustainable development project, or just the same old corporate incentives in trendy new clean energy bottles" (*http://www.austinchronicle.com/issues*, accessed January 27, 2006). The city council and Austin Energy are also the driving force behind a coalition of U.S. city governments, nongovernmental organizations, and utility companies[2] established to lobby auto manufacturers to produce plug-in hybrid gasoline–electric engines and have pledged to purchase 600 of these for the city's own municipal fleet (*http://www.msnbc.msn.com/id/10990145*, accessed February 2, 2006).

THE "BOSTON FORMULA" FOR SUCCESS

Boston, Massachusetts, is the "hub" of one of the United States' most competitive economic regions. In its approximately 350-year history

Boston has responded to changes in the structure of the global economy, from mercantilism to neoliberalism. Until the 1930s, for example, the region was known for textiles and manufacturing. In the late 1960s, from the strength of Raytheon and other defense contractors, the Boston metropolitan area evolved into a signature region of the military production complex, both in terms of innovation and manufacturing. After losing the battle over the computer industry to Silicon Valley, the city and regional economy waned again in the late 1980s. The recession that followed was the worst in the region since the Great Depression of the 1930s. From 1970 to 2000 the region lost 35,000 manufacturing jobs, bringing the percentage of those jobs from 12% of the region's employment down to 4% (Boston Foundation, 2004). During this same time Boston's knowledge economy grew to some 68% of employment (Boston Foundation, 2004). The Boston region is now widely recognized as one of the world's most innovative economies. Boston's shift from a manufacturing economy to a knowledge-based economy is rooted in the research and development emerging from the area's institutions of higher education and health care, as well as the financial, governmental, business, professional, and human services sectors. By many indicators Boston ranks as one of the most competitive city-regions in the world. The Boston area's population is highly educated, with 34% of adults holding a bachelor's degree (Boston Foundation, 2004). According to the Massachusetts Technology Collaborative, Boston also produces patents at a rate of 61 annually for every 100,000 residents. This is higher than in any of the region's U.S. competitors. In terms of R&D, Boston is outpaced only by Silicon Valley for venture capital access and federal research funding. The Boston metro area houses half of the state's population and jobs. The city of Boston alone has 9% of the population and 16% of the state's jobs. During the growth period of the 1990s the Boston area's population increased at a rate of 5.5%. This overall modest rate conceals the explosive population growth of some towns in the region. Chelsea's population, a town to Boston's immediate north, increased by 22%, while cities around the emerging high-tech corridor of Interstate 495 grew at a pace of 45% during the same time period. Overall, while the city of Boston gained 5.5%, the high-tech communities to the west grew at a rate of 11.3% during the same time. Boston's economy generates substantial commuter traffic; while the city's total population is about 600,000, the number of people in the city doubles each day during working hours, some 300,000 workers making their way into the city via rail or automobile.

In a similar fashion to Austin, the city's administration has pursued several green initiatives. In 1995, for example, the Office of the Mayor initiated Sustainable Boston, which was modeled after Local Agenda 21, was supervised by ICLEI, and included an indicators project. In 1995 and 1996 public forums were held around the city to begin a discussion on the city's quality of life issues. Over 2,000 citizens attended visioning sessions, which were intended to produce a common vision for a sustainable Boston. A report from this effort, *The Wisdom of Our Choices: Boston's Indicators of Progress, Change and Sustainability*, was published by the Boston Foundation in 2000. Before the report was published, however, the Sustainable Boston Initiative floundered. Portney (2003: 224) reports that the initiative "quickly took a back seat to traditional economic development in the city's priorities." This supports Lake's (2000) observation that the city of Boston was motivated less by the principles of sustainability and more by the potential for stimulating economic growth and urban redevelopment.

While the successor plan, Boston 400, retained some of the influences of the Sustainable Boston plan, it was decidedly more focused on economic development. The Boston 400 plan was housed in the Boston Redevelopment Authority, the city's primary economic development agency, whereas Sustainable Boston had been located in the city's environment department, primarily charged with environmental compliance and monitoring. In his remarks to the American Planning Association in 1998 the city's chief economic development officer Thomas O'Brien revealed the economic development-oriented nature of Sustainable Boston's successor. He argued that "economic vitality leads to opportunity, as projects which were once only ideas can now become a reality." A key component of this plan was the commitment of $700 million by the city to waterfront redevelopment to improve tourist and commercial opportunities in east Boston, based on the premise that by planning for, and promoting, the right set of assets, such as Boston's waterfront, the city as a whole would benefit.

Yet, while economic development is the cornerstone of prosperity, the Boston Redevelopment Authority (BRA) also recognized a broader set of city assets. Haar (1998, p. 1), while describing the plan, remarked:

> Boston's primary assets are the mixed-use character of our neighborhoods, a multi-faceted economy, outstanding locational advantages for business and industry firms, world-renowned educational and medical

institutions, a strong community service system, and a rich and diverse cultural and ethnic heritage, not to mention the City's vibrant and walkable urban fabric and its location near one of the most attractive harbors anywhere. . . . Boston 400 must strengthen the City's already firm, reliable infrastructure to enhance a wide range of activities—so that everyone in the City can respond to challenges that we cannot predict today.

As a comprehensive plan, however, the Boston 400 plan, like its predecessor, Sustainable Boston, withered on the political vine. As one former employee of the BRA (interview, June 2006) put it, "The Mayor wasn't completely behind it, so we had no incentive to keep it going." Aspects of both plans remain alive today both discursively and materially, though in a more piecemeal way and spread out through several city departments and local growth coalitions. Discursively, it would seem that the broader set of issues identified in Sustainable Boston and Boston 400 have been codified in what the Boston Foundation, among others, has referred to as the "Boston Formula" (see Table 4.2). Implied here, and explicit in the Boston Foundation's publication *Thinking Globally/ Acting Locally: A Regional Wakeup Call* (2004), is that Boston is in competition with other high-tech areas such as Chapel Hill, North Carolina, and the San Francisco Bay area and that these competitor "city-states" are applying similar formulae to attract workers to their area. Thus, Boston's quality-of-life assets, such as its Olmstead-inspired "Emerald Necklace," its waterfront and other open spaces, and the New England village image, in addition to economic factors, are crucial to the success of the region in the new economy.

Other initiatives also represent the city's efforts to deploy the sustainability discourse to promote itself in today's entrepreneurial envi-

TABLE 4.2. The Boston Formula for Success

- Excellent higher education institutions with large student bodies, including foreign students skilled in math and science.
- Highly educated young workers and families.
- Access to private venture capital and public research funding.
- Culturally vibrant and walkable neighborhoods.
- City and town centers near public transit.
- Nearby recreational and natural areas.

Note. Data from the Boston Foundation (2004; emphasis added).

ronment. In the fall of 2005 the city had unveiled its own green roof atop Boston City Hall. At the ribbon cutting of the new roof the Mayor announced, "I am determined to make the city of Boston a leader in green technology. Not only will it keep us on the cutting edge; it also just makes good sense—for our budgets and for our environment." In May 2006 the Boston Environment Department hosted a green roofs conference. Another of Boston's environmental initiatives is the city's groundwater overlay districts, first established in 1986. In 2005 Mayor Thomas M. Menino announced that the whole city would become part of the groundwater overlay district. The development of the district also marks a new watermark in collaboration between the city, the state, and local nonprofit organizations. New projects that fall within the district (the actual borders are being shaped by the political process) will undergo a review that measures their implications for groundwater levels. The event that set this initiative into motion was not water scarcity per se but rather the physical settling experienced in many of Boston's famous historic structures, which are seen as a crucial element in Boston's formula for success in the new economy. The wooden pillars supporting many of Boston's famous historic brownstones were being exposed to air and bacteria as the water table gradually dropped from overconsumption generally.

As well as being evident in various piecemeal city efforts over the past decade, Boston's original sustainability initiative and its successors have also been manifest in regional and state activities. In 2002, the Metropolitan Area Planning Council (MAPC), the regional planning entity for Boston and its 100 surrounding towns, initiated the "Metro-Future" project. MetroFuture is designed to update the earlier Metroplan developed in the early 1990s, which was the region's first effort to grapple with regional rather than town planning. The need for the plan is being driven by a familiar theme. According to proponents of the plan:

> We live in a very desirable region. The quality of life offered by our historical, cultural, natural, and economic attributes continues to retain residents and draw increasing numbers of people to make their homes here. As the number of households in Metropolitan Boston is rising, we are all—long-time resident and newcomer alike—placing increasing demands on the infrastructure that supports our quality of life. Indeed, we appear to be jeopardizing many of the attributes that drew us here in the first place. (Metropolitan Area Planning Council, 2002)

The goal is to develop a sustainable growth plan for the MAPC region, including implementation steps for state and local government and recommendations for private sector stakeholders. To do so, the MAPC is following a Local Agenda 21 style process, involving large-scale public education to increase the visibility and awareness of regional issues related to the economy, environment, and quality of life; region-wide civic engagement in the planning process, helping to build a constituency of knowledgeable and committed supporters who will work to translate the plan into reality; and institutional capacity building throughout the region, linking technology and information to community decision making for current and future planning processes. This MetroFuture effort is ongoing and at the time of writing the team was in the midst of stage 2 of the plan. The rhetoric behind the plan reflects the historical trajectory of sustainability in the region over the past decade. The Local Agenda 21 style of visioning, consensus building, and implementation, the link between economic competition and sustainability and environmental concerns, and the relative absence of business from the process all reflect Boston's previous experiences in sustainability.

The manifestation of sustainability and economic competitiveness as public policy and planning concerns are also present at the state level. The Commonwealth of Massachusetts, through the efforts of the Office of Commonwealth Development (OCD) and the state legislature, have attempted to address sustainability concerns and economic competitiveness through the state's most pressing public crisis, affordable housing. There is not enough space to go into detail here, but a brief discussion is necessary. In 2005 the state legislature passed amendments to the affordable housing statute that included certain "smart-growth" components. In particular, the state would provide incentives for developments that used existing developed sites (not open space) and were near rail transport terminals. The statute further required that 20% of housing at these sites had to be "affordable." As a further incentive to cities and towns across the state, the Office of Commonwealth Development linked its funding allocations (nearly $500 million annually) to its sustainable development checklist. The checklist would be evaluated and weighted into grant applications submitted to the state. What this means is that all cities and towns that seek state funding for road or development projects must (1) complete the form and (2) rank highly enough on its criteria (or at least higher than other towns) to receive state funding.

GOVERNING SUSTAINABLE NEW ECONOMY SPACES?

These two case studies reveal, perhaps not surprisingly, that new economy spaces are probably as diverse as the landscapes they occupy. A concern for sustainable development certainly exists alongside the new economy in both locations, but it is questionable as to whether this really represents a shift toward greater "ecological rationality." For example, rapid growth in the Boston region and in Austin has affected their respective groundwater supplies. While adequate water supplies are obviously crucial, the issue of scarcity has been linked less to future consumption than to the prospective loss of future amenities. In Austin, Barton Creek is a regional icon: not only does it provide drinking water for the region's inhabitants, but it also provides the backdrop for many of the region's recreational amenities. Thus, its importance to the region as both a water supply and amenity is clear even to firms seeking to expand their operations in the area. Boston's historic "Back Bay" is similar to Barton Creek in cultural significance. Built 140 years ago, the Back Bay neighborhood is constructed on landfill that was placed over the Charles River estuary. Not only does this historic landmark area house Boston's "creative class," but also it attracts tourists, and the wide tree-lined sidewalks provide a resource to residents and visitors alike.

In both locations, concerns over equity have tended to take a backseat to concerns over quality of life issues as factors in local economic competitiveness. In both cases, smart-growth initiatives have formed a mainstay of attempts to address these concerns. In Austin these initiatives have sought to overcome the impasse over development, while in Boston one concern has been to maintain the distinctiveness of both the "Boston formula" and the New England village character felt to be crucial in the city's appeal to the creative class. However, although Austin may have smart-growth and neighborhood planning initiatives, these exist alongside substantial income disparities and concerns that "smart growth" is encouraging gentrification in low-income inner-city neighborhoods (McCann, 2003). In Boston, regional and state efforts to promote smart growth have resulted mainly from the soaring property values in the region. Boston is more expensive to live in than Austin, making it the most expensive rental market in the country. Yet, for policymakers this is only part of the problem. There is a growing concern by young people from the region that they will be unable to afford to live in the region after they are graduated from college. Furthermore, as commercial development in the Boston city-region expands outward with its associated land-use demands, it impacts upon the space available for workers to live in and to get to work. This concern lies behind the state's incentives for Transport Oriented Development (TOD)—high-density development along rail and commuter links. The impetus behind smart growth in Boston is thus to retain and attract key workers for the knowledge economy and less about housing opportunities for those who participate in the ancillary economies.

In both of the case studies the two cities certainly face similar new economy problems. In both, traffic congestion, the environmental consequences of rapid growth, high property values, income disparities, and increasingly scarce natural resources are all major concerns. However, the form of concern for sustainability and quality of life issues differs between the two areas, as have the forms of institutional response. Austin's governing coalition has been able to forge alliances between NGOs, environmental groups, new economy firms, and the state. New forms of governance have emerged in Austin with the creation of public–private partnerships, where issues of sustainability have been addressed as a factor in maintaining economic competitiveness through quality of life issues but also as a potential future accumulation strategy—for example, through promoting the clean energy sector as a source of future growth. In Austin, there has been much greater corporate involvement in sustainability initiatives than in Boston, where the public sector has struggled with the lack of corporate engagement. One explanation may be that whereas in the (Fordist) past Boston had two or three key firms that were civic minded and participated in local politics beyond their self-interest, over the past decade these companies have been sold to larger ones and their headquarters have moved out of the city or have closed altogether. As a result the city of Boston has not been able to create a strong culture of participation in land-use or development affairs outside the typical pathways. From our interviews with planners involved in the MetroFutures project, getting business firms to the table has been one of the biggest obstacles in the process. In contrast, at the subregional level along the 128 and 495 high-tech beltways, new economy firms are participating, which is more in line with the Austin experience. Rather than reaching any form of consensus (forced or otherwise), development politics in Boston remains clannish. While Austin's new economy firms came to the table and worked to help protect environmental resources (regardless of motivation), Boston firms (either in the city or around the metropolitan area) are largely absent from planning. Indeed, as one respondent in the Office of Commonwealth Development noted, corporate involvement is largely restricted to "firms telling the governor what they want in terms of quality of life and infrastructure" (interview, OCD, June 2006). They do not, however, work with the Boston local authority to resolve problems through the planning process. Moreover, our respondents suggested that new economy workers in Boston participate even less than their employers. According to one respondent, "New economy workers in Boston tend to be rather young and not have children." He feels that, as a result, they are not as closely tied to the community. "They are motivated to go out and protest global warming or the war in Iraq, but the mundane issues of local development don't motivate them to attend a planning meeting" (interview, Boston Environment Department, June 2006). Thus, as previous scholars have noted (Saxenian, 1994; Horan & Jonas, 1998) the "Massachusetts miracle" that propelled the region's economy during the 1980s hardly constituted a virtuous form of local or regional governance. Saxenian (1994) observed that Boston was overhierarchical, reflecting a disposition toward rigidity and not enough flexibility and collaboration in the corporate gene pool. Apparently, this condition has worsened in wake of the recent sale of such local companies as Converse, Gillette, and John Hancock. As one official from Boston's Environment Department (BED) put it, "Firms such as Bank of Boston or Very Fine Juice were willing to put up something; now there is no clear place to go to support local initiatives" (interview, BED, May 2006).

CONCLUSIONS

In this chapter we began with a number of key questions to be addressed: Do material conditions in new economy spaces support both a model of global competitiveness and the principles and practices of sustainable development? Is a concern for sustainability integral to the development of these new economy spaces? What policy forms does a concern for sustainability take in these localities? Finally, what institutional forms have evolved in new economy spaces to address these issues?

    With regard to the first three questions, it appears that new economy spaces set out on "sustainable" pathways when the material conditions that underpin them are compromised by rapid growth. In the spaces of the new economy there are two major problems that need to be addressed. First, there is a contradiction between the economic growth of such areas and quality of life/"good environment." While the success of these areas is intimately bound up with quality of life issues for the elite groups of high-tech workers, the consequences of growth through congestion, housing costs, and sprawl may make these areas less attractive to highly mobile workers. The second problematic issue is over the inequitable impacts of growth and prospects for continuing success. Those lower-paid workers who service the new economy are increasingly unable to afford the costs of housing in such areas and may suffer from poor housing, long commuting times, and poor working conditions. Thus, in the case of Austin, engagement with the sustainable development agenda has been framed around preserving the reason many knowledge workers come to the area—its natural environment. Boston is similar, but here the issue of quality of life is problematized by the scarcity of affordable housing. What both cases suggest, however, is that there is a certain aesthetic about these places that must be preserved, whether it be the open space, historic buildings, the "urban village," or the outdoor lifestyle. In this sense then it would appear that sustainable development is a key component that is mobilized in support of existing economic trajectories. In the spaces of the new economy, economic success, quality of life, and a "good environment" are closely intertwined. It would appear, prima facie, that the economic competitiveness of new industrial spaces is contingent upon continuing engagement in the sustainability discourse. However, at a more theoretical level, we cannot assume that a greater concern for environmental issues in new economy spaces simply represents sustainability in practice. While, at least in its "stronger" forms, sustainable development may represent a political challenge and an alternative to "development as usual," the concern for continued economic success may simply see the environment as a means of securing accumulation regardless of the social costs involved. Thus, in new economy spaces as elsewhere, it is possible that "the current proliferation of sustainability projects and products represents little more than a strategy to secure conditions for the continuance of accumulation-as-usual" (MacBride, 2004: 341). As the ideology of neoliberalism continues to hold sway, economic decision making increasingly dominates the political agenda and thus maps directly onto the sustainability agenda. Because of this inescapable engagement with capitalist social relations, the true intent of "sustainable development policies" is frequently marginalized (Gibbs, Jonas, & While, 2002; Lake, 2000).

Sustainable development has thus been deployed as an alternative set of policy options, or institutional fixes, in cities seeking to curb environmental transformation and degradation from rapid economic development. Indeed, those high-tech U.S. cities that top the index for competitiveness are the same cities that also rank most highly on the key metric of sustainable development. The claim here is not that these cities are more sustainable, as city boosters would have us believe. Rather, it is that they have adopted the sustainable development discourse and formulated policy options to address the tensions from their models of growth and economic competition. The types of sustainable development institutions and policy proscriptions that emerge are partially a result of the constraints of a neoliberal discourse. That is, they are crafted hybrids of a market-based ideology—incentives, volunteerism, and private sector rationality carry the day. After all, whether "rolling back" the state or "rolling out" neoliberalism, the recent political power of economic liberals has fundamentally transformed state–society relations (Peck & Tickell, 2002; Raco, 2005). The current harmony between economically liberal social conservatives and third-way progressives will become cacophonic at some point. Perhaps more important is the observation that the politics surrounding regulation today is not *all* of a neoliberal orientation. Actors engaged in the politics of regulatory change represent many differing views. The resulting institutions and policy choices do not necessarily follow the straight edge of the neoliberal project. Contingency solutions potentially exist when actors struggle to define just what the problem is that institutions must ameliorate.

NOTES

1. A total of five interviews were conducted in Austin in 2003 as part of an Economic and Social Research Council funded project (Grant No. R000239428) and eight interviews in the Boston area in 2006.
2. The Plug-in Partners Coalition also involves Arlington (VA), Baltimore, Boston, Fort Worth, Corpus Christi, Denver, Irvine (CA), Los Angeles, Seattle, and Wenatchee (WA) (see *www.pluginpartners.org*).

REFERENCES

Agyeman, J. (2005). *Sustainable communities and the challenge of environmental justice*. New York: NYU Press.
Agyeman, J., Bullard, R., & Evans, B. (2003). *Just sustainabilities: Development in an unequal world*. Cambridge, MA: MIT Press.

Allen, J., Massey, D., & Cochrane, A. (1998). *Rethinking the region*. London: Routledge.

Angel, D. (2000). Environmental innovation and regulation. In G. L. Clark, M. P. Feldman, & M. S. Gertler (Eds.), *The Oxford handbook of economic geography* (pp. 607–622). Oxford, UK: Blackwell.

Atkinson, R. (1999). *The state new economy index*. Washington, DC: Progressive Policy Institute. Available at *http://www.neweconomyindex.org/states/1999/introduction.html*.

Atkinson, R. (2002). *The state new economy index*. Washington, DC: Progressive Policy Institute.

Beatley, T. (2000). *Green urbanism*. Washington, DC: Island Press.

Bishop, B. (2000). Austin's new economy outpacing its old government. Accessed June 22, 2005, at *http://www.citistates.com/essays/essay_neweconomy.html*.

Bluestone, B. (2006). *Sustaining the mass economy: Housing costs, population dynamics and employment*, Center for Urban and Regional Policy, Northeastern University. Available at *http://www.curp.neu.edu/publications/reports.htm#sustaineconomy*.

Boston Foundation. (2000). *The wisdom of our choices: Boston's indicators of progress, change and sustainability*. Boston: Boston Foundation.

Boston Foundation. (2004). *Thinking globally/acting locally: A regional wakeup call*. Boston: Author.

Breheny, M. (1995). The compact city and transport energy consumption. *Transactions of the Institute of British Geographers, 20*(1), 81–101.

Buckingham, S., & Lievesley, G. (2006). *In the hands of women: Paradigms of citizenship*. Manchester, UK: Manchester University Press.

Buckingham, S., & Theobald, K. (2003). *Local environmental sustainability*. Cambridge, UK: Woodhead Press.

City of Austin (2001). *Smart growth initiative: Matrix application packet*. Austin: Austin Transportation and Design Department.

Cox, K., & Jonas, A. E. G. (1993). Urban development, collective consumption and the politics of metropolitan fragmentation, *Political Geography, 12*(1), 8–37.

Cox, K., & Mair, A. (1988). Locality and community in the politics of local economic development. *Annals of the Association of American Geographers, 88*, 307–325.

Crang, P., & Martin, R. (1991). Mrs. Thatcher's vision of the "New Britain" and the other sides of the "Cambridge Phenomenon." *Society and Space, 9*, 91–116.

Daniels, P., Beaverstock, J. V., Bradshaw, M. J., & Leyshon, A. (2005). *Geographies of the new economy*. London: Routledge.

Florida, R. (2002). *The rise of the creative class: And how it's transforming work, leisure, community, and everyday life*. New York: Basic Books.

Gibbs, D. (2002). *Local development and the environment*. London: Routledge.

Gibbs, D., Deutz, P., & Proctor, A. (2005). Industrial ecology and eco-industrial development: A new paradigm for local and regional development? *Regional Studies, 39*(2), 171–184.

Gibbs, D., & Jonas, A. (2000). Governance and regulation in local environmental policy: The utility of a regime approach. *Geoforum, 31*, 299–313.

Gibbs, D., Jonas, A., & While, A. (2002). Changing governance structures and the environment: Economy–environment relations at local scales. *Journal of Environmental Policy and Planning, 4*, 123–138.

Gibbs, D., & Krueger, R. (2004, March 15–20). *Toward theorising a political economy of sustainability*. Paper presented to the session "Theorizing Sustainability: Connecting Economy, Environment, and Society" at the Association of American Geographers annual meeting, Philadelphia.

Gleeson, B., & Low, N. (2000). Cities as consumers of the world's environment. In N. Low & B. Gleeson, (Eds.), *Consuming cities: The urban environment in the global economy after the Rio Declaration* (pp. 1–29). London: Routledge.

Gottlieb, P. (1995). Residential amenities, firm location and economic development. *Urban Studies, 32*(9), 1413–1436.

Greater Austin Chamber of Commerce. (2005). *Customized report*. Accessed January 27, 2006, at *http://www.austinchamber.com*.

Gunn, K. (2004). *Exploring environmental policy in Austin, Texas*, unpublished Masters of Public Administration thesis, Texas State University, San Marcos.

Haar, L. (1998). *Boston 400: Shaping the future city*. Proceedings of the 1998 National Planning Conference, Tempe, AZ: American Institute of Certified Planners Press.

Harvey, D. (1985). *The urbanization of capital: Studies in the history and theory of capitalist urbanization*. Baltimore: Johns Hopkins University Press.

Haughton, G. (1999). Environmental justice and the sustainable city. *Journal of Planning Education and Research, 18*, 233–243.

Hempel, L. C. (1999). Conceptual and analytical challenges in building sustainable communities. In D. A. Mazmanian & M. E. Kraft (Eds.), *Towards sustainable communities: Transition and transformations in environmental policy*. Cambridge, MA: MIT Press.

Her Majesty's Government. (2005). *Securing the future: The U.K. government sustainable development strategy*. Cm 6467, Norwich, UK: The Stationery Office.

Herrshel, T., & Newman, P. (2002). *Governance of Europe's city regions: Planning, policy and politics*. London: Routledge.

Herzog, H., & Schlottman, A. (1991). Metropolitan dimensions of high-technology location in the U.S.: Worker mobility and residence choice. In H. Herzog & A. Schlottman (Eds.), *Industrial location and public policy*. Knoxville: University of Tennessee Press.

Horan, C., & Jonas, A. (1998). Governing Massachusetts: Uneven development and politics in metropolitan Boston. *Economic Geography*, Extra Issue, 83–95.

International Council for Local Environmental Initiatives. (2002). *Second Local Agenda 21 survey*. New York: United Nations Commission on Sustainable Development.

Jessop, B. (1990). *State theory: Putting capitalist societies in their place*. Cambridge, MA: Polity Press.

Jessop, B. (1995). The regulation approach, governance and post-Fordism: Alternative perspectives on economic and political change. *Economy and Society, 24*, 307–333.

Jessop, B. (2002). *The future of the capitalist state*. Cambridge, MA: Polity Press.

Jonas, A. E. J., Gibbs, D., & While, A. (2004). State modernisation and local strategic selectivity after Local Agenda 21: Evidence from three northern English localities. *Policy and Politics, 32*(2), 151–168.

Judd, D. R. (1999). Constructing the tourist bubble. In D. R. Judd & S. S. Fainstein (Eds.), *The tourist city* (pp. 35–53). New Haven, CT: Yale University Press.

Kelly, K. (1998). *New rules for the new economy*. London: Fourth Estate.

Kong, L. (2000). Culture, economy, policy: Trends and developments. *Geoforum, 31*, 385–390.

Krueger, R. (2002). Relocating regulation in Montana's gold mining industry. *Environment and Planning A, 34*, 867–881.

Krueger, R., & Agyeman, J. (2005). Sustainability schizophrenia or "actually existing sustainabilities"?: Toward a broader understanding of the politics and promise of urban sustainability in the U.S. *Geoforum, 34*, 410–417.

Krueger, R., & Buckingham, S. (2005). *Painting the town green: Using a critical sustainable development framework to evaluate the "creative city" discourse*. Sustainable Cities Research Group Paper Series, SD: 002.

Lake, R. (2000). Contradictions at the local state: Local implementation of the U.S. sustainability agenda in the USA. In B. Gleeson & N. Low (Eds.), *Consuming cities: The urban environment in the global economy after the Rio Declaration* (pp. 70–90). London: Routledge.

Landry, C. (2000). *The creative city*. London: Earthscan.

Ley, D. (1996). Urban geography and cultural studies. *Urban Geography, 17*(6), 475–477.

Luke, T. (1996). *Ecocritique*. Minneapolis: University of Minnesota Press.

Luke, T. W. (2003). On the political economy of Clayoquot Sound: The uneasy transition from extractive to attractive models of development. In W. Magnusson & K. Shaw (Eds.), *A political space: Reading the global through Clayoquot Sound* (pp. 91–112). Minneapolis: University of Minnesota Press.

MacBride, S. (2004). Production and the revenge of nature: Material transformations in Hudson's *Producing Places*. *Antipode, 36*(2), 337–343.

McCann, E. J. (2003). Framing space and time in the city: Urban policy and the politics of spatial and temporal scale. *Journal of Urban Affairs, 25*(2), 159–178.

McCann, E. J. (2005). Inequality and politics in the creative city-region: questions of livability and state strategy. *International Journal of Urban and Regional Research, 31*(1), 188–196.

Metropolitan Area Planning Council. (2002). *Metrofuture Plan for a Greater Boston Region*. Accessed April 17, 2007, at *http://www. metrofuture.org*.

Molotch, H. L. (1996). L.A. as product. In A. Scott & E. Soja (Eds.), *Los Angeles and urban theory at the end of the twentieth century* (pp. 225–277). Berkeley: University of California Press.

Neveraz, L. (2003). *New money/nice town: How capital works in the new urban economy*. New York: Routledge.

Nichols, I. (2006, October 6). More TODs to come. *Austin Chronicle*. Accessed October 24, 2006, at *http://www.austinchronicle.com*.

O'Connor, J. (1998). *Natural causes: Essays in ecological Marxism*. New York: Guilford Press.

O'Riordan, T. (1999). *Planning for sustainable development*. Tomorrow Series. London: The Town and Country Planning Association.

Opportunity Austin. (2005). *A solid platform for a solid future* (annual report). Austin, TX: Greater Austin Chamber of Commerce.

Peck, J., & Tickell, A. (2002). Neoliberalizing space. *Antipode, 34*(3), 380–404.

Pellow, D. N., & Park, L. S. (2002). *The Silicon valley of dreams: Environmental injustice, immigrant workers, and the high-tech global economy*. New York: New York University Press.

Platt, R. (2004). *Land use and society*. Washington, DC: Island Press.

Porter, M. (1990). *The competitive advantage of nations*. Cambridge, MA: Harvard University Press.

Portney, K. (2003). *Taking sustainable cities seriously: Economic development, quality of life, and the environment in American cities.* Cambridge, MA: MIT Press.

Prytherch, D. (2002). Selling the eco-entrepreneurial city: Natural wonders and urban strategems in Tucson, Arizona. *Urban Geography, 23*(8), 771–793.

Raco, M. (2005). Sustainable development, rolled-out neoliberalism and sustainable communities. *Antipode, 37*(2), 325–347.

Sassen, S. (1991). *The global city.* Princeton, NJ: Princeton University Press.

Saxenian, A. L. (1994). *Regional advantage: Culture and competition in Silicon valley and route 128.* Cambridge, MA: Harvard University Press.

Scott, A. J. (1997). The cultural economy of cities. *International Journal of Urban and Regional Research, 21*(2), 323–339.

Scott, A. J. (2000). *The cultural economy of cities.* Beverly Hills, CA: Sage.

Scott, A. J. (2001). *Global city regions: Trends, theory, policy.* Oxford, UK: Oxford University Press.

Segal, Quince and Wicksteed. (2000). *The Cambridge phenomenon revisited.* Cambridge, UK: Segal, Quince and Wicksteed.

Smilor, R. W., Gibson, D. V., & Kozmetsky, G. (1988). Creating the technopolis: High-technology development in Austin. *Journal of Business Venturing, 4*, 49–67.

Storper, M. (1997). *The regional world: Territorial development in a regional world.* New York: Guilford Press.

Thrift, N., & Olds, K. (1996). Refiguring the economic in economic geography. *Progress in Human Geography, 20*(3), 311–337.

Tinker, A. K. (2003). *The Austin green building program: An analysis of the program's effectiveness,* unpublished PhD thesis, Texas A&M University, College Station, TX.

Wackernagel, M., & Rees, W. (1996). *Our ecological footprint: Reducing human impact on the earth.* Gabriola Island, BC, and New Haven, CT: New Society Publishers.

Walker, P. A. (2003). Reconsidering "regional" political ecologies: Towards a political ecology of the rural American West. *Progress in Human Geography, 27*, 7–24.

Watson, M. (2001). Embedding the "new economy" in Europe: A study in the institutional specificities of knowledge-based growth. *Economy and Society, 30*, 504–523.

While, A., Jonas, A., & Gibbs, D. (2004). Unblocking the city? Growth pressures, collective provision, and the search for new spaces of governance in Greater Cambridge, England. *Environment and Planning A, 36*, 279–304.

World Commission on Environment and Development. (1987). *Our common future.* Oxford, UK: Oxford University Press.

Zukin, S. (1995). *The cultures of cities.* Oxford, UK and Cambridge, MA: Blackwell.

CHAPTER 5

Greening the Entrepreneurial City?

Looking for Spaces of Sustainability
Politics in the Competitive City

ANDREW E. G. JONAS
AIDAN WHILE

> We must understand that the cities which wreak the most havoc
> [on the environment] are mainly those in the West, both in their
> own right and as a model which everyone seeks to imitate. . . .
> We can find an instance of this at home. Barcelona and its
> metropolitan area have experienced hardly any population growth
> since the seventies. However, land occupation and water and
> energy consumption have increased exponentially. This is the
> model which needs to be put right. . . . We must learn to do
> things another way, no matter how much powerful networks of
> interest systematically oppose the introduction of modifications
> which are indispensable to our model of society.
> — JOSEP RAMONEDA, DIRECTOR OF THE CENTRE DE CULTURA
> CONTEMPOÀNIA DE BARCELONA (CCCB, 1998: 8)

This quotation from the Director of the Centre de Cultura
Contemporània de Barcelona, Josep Ramoneda, captures in many respects the contradictions of the new sustainability imperative confronting urban governance. In recent years, urban political leaders in the West
have devoted considerable time and public resources to the pursuit of

123

various strategies of postindustrial urban economic development. At the same time, they have attempted to confront growing social inequalities among urban residents and increasing demands to protect local, regional, and global environmental resources. Many urban political leaders now see entrepreneurialism and sustainability as not necessarily in conflict but instead as compatible strategic political and policy goals; indeed, promoting environmental and social sustainability is increasingly seen as exemplary urban governance. Given this, it is perhaps a paradox to find that many theoreticians of contemporary urban politics have examined the entrepreneurial city separately from the sustainable city, implying that these represent theoretically incommensurable ideal types.

In this chapter, we attempt to identify points of contradiction and commensurability in the literatures on urban entrepreneurialism and sustainability, respectively and together. After a cursory review of these literatures, we offer a short case study of an urban-scaled politics of social and environmental sustainability in Barcelona, the capital of the Spanish provincial region of Catalunya. Why Barcelona? For many academics and policymakers, the city of Barcelona represents a model of economic, social, and political transformation (Marshall, 2004). Yet, it is not always seen in the same light as an exemplar of good urban governance for the environment. This apparent failure on the part of urban leadership to meet the challenges of social and environmental sustainability is a point conceded by even the most faithful defenders of the city's bold transformation model. Nonetheless, recounting elements of the Barcelona story allows us to argue that there is increasing evidence from so-called competitive cities that urban leaders *are* beginning to respond to various political demands to be more accountable—or less unaccountable—for the ecological footprint of their cities and citizens. More than simply a matter of greening urban space and mitigating the environmental damage caused by urban growth, entrepreneurial cities are developing sophisticated governance structures and political strategies to incorporate competing economic, social, and ecological demands.[1] Yet, at the same time in cultural and economic centers such as Barcelona tensions around social and environmental sustainability remain, which open up possibilities for policy makers, political actors, and activists to develop ecological strategies and social policy alternatives to those of urban competitiveness and place promotion (i.e., neoliberal urbanism). It is how to approach and explain the diverse nature and complexity of such emerging spaces of sustainability politics in the Western city that inspires this chapter.

We anticipate that our chapter will raise more questions for critical urban research than we can possibly answer here. To what extent under neoliberalism has the urban become a *necessary* site and scale of sustainability politics? Do the practices and politics of urban sustainability offer alternatives to those of neoliberal urbanism? How should critical urban research approach empirically the (often contradictory) ways in which economic, social, and environmental actors in entrepreneurial cities and regions engage with governance processes and develop political strategies? What new spaces of political activism, and of state and governance, are emerging around urban sustainability? These, then, are some of the questions we intend to provoke in this chapter.

One of the arguments we develop is that sustainability is often conflated with environment or ecology, thereby obscuring the social dimension. The chapter attempts to move beyond such a conflation and considers how perhaps less tangible social dimensions of urban sustainability, such as social inclusion, cultural diversity, health, nutrition, accessibility, transport, mobility, housing, and community development, are expressed alongside, or perhaps even in conflict with, ecological imperatives. At the heart of this inquiry is a need to investigate how the variety of social, environmental, and economic interests in the entrepreneurial city are expressed in particular strategies, projects, and policies relating to sustainability, and to reveal the changing nature of territorial, class, and political alliances associated with these.

ENTREPRENEURIALISM, SUSTAINABILITY, AND URBAN GOVERNANCE

Urban Entrepreneurialism

An almost taken-for-granted axiom of Western critical urban theory is that today's cities are embedded in global circuits of capital, and that in order to compete successfully urban places must increase and intensify their connections to such circuits. Almost two decades ago, Harvey (1989) characterized this phenomenon as a new era of urban entrepreneurialism and drew a comparison with the previous period of urban managerialism. Urban elites were, he suggested, engaged in a desperate search for some sort of new economic and social role for their cities in wider spatial divisions of production and consumption. This could be achieved perhaps by a city's becoming a financial command and control center, a cultural capital, the headquarters location of major international corporations and governmental organizations, and so forth. In the process, it was argued that urban politicians were forging new partnerships and institutional relations with business elites and representatives of international capital, and old alliances and coalitions of a more redistributive-liberal or radical-leftist character were fragmenting or disappearing altogether. Under entrepreneurialism, cities and their spaces of politics were becoming neoliberalized (Peck & Tickell, 2002); that is to say, urban governance reflected the dictates of the global marketplace more than the will and needs of, for instance, the urban electorate.

As urban scholars caught on to this trend (see also Judd & Ready, 1986), new theoretical narratives of urban entrepreneurialism emerged alongside more established urban political theories, including those of the urban growth machine (Logan & Molotch, 1987) and urban regime (Stone, 1989, 1993). These theories in turn were revisited and reconstructed to fit with the new political and economic times (see, for example, contributions in Hall & Hubbard, 1998; Jonas & Wilson, 1999; Lauria, 1997). Each of these reconstructed approaches, in its own way, has been dealing with a tendency common to all cities: the appearance of the "new urban politics" (Cox, 1993). For the most part, these politics are concerned with economic development and are infused with an ideology of "growth first" or "growth at all costs." Urban-based economic and political actors have mobilized around new territorial (rather than, say, traditional leftist or class-based) alliances across the city and have undertaken various urban place promotion strategies, transformations in the built environment, and urban governance reforms so as to intensify the "globalness" of a city. Other political goals of a perhaps more redistributive intent, along with the specific interests concerned with these goals, have as a consequence been marginalized or excluded (Cox & Mair, 1988).

Furthermore, new strategic capacities and institutional arrangements have been rolled out by the state to cope with globalization and associated processes of urban economic and social restructuring. To some it seems that the state has been steering urban managers to adopt regeneration strategies and policies, the effect of which is to encourage cities to compete for ever more scarce resources (Jessop, Peck, & Tickell, 1999). Although reconstructed urban theories allow some scope for studying resistance on the part of urban publics (Brenner & Theodore, 2002), the emerging consensus is that *all* cities are becoming entrepreneurial. Differences between cities in terms of politics and policy outcomes are matters of degree rather than substance (Jessop et al., 1999).

If anything, such differences reflect different national priorities rather than arising from the expression of local political interests.

The Sustainable City

Nonetheless, some urban scholars have remained remarkably unmoved by all the recent attention to urban entrepreneurialism. These note that certain forms of urban governance do not fit a dominant Western (especially "American") capitalist model of the entrepreneurial city (see Jonas & Wilson, 1999). Not all urban strategies have followed a well-trodden path of cities undergoing postindustrial transition and engaging in a desperate search for mobile capital, jobs, investments, and grants. To some extent this is not surprising in that many of the world's major cities are not in fact in or of the "West" and that, consequently, perhaps too much has been read into transformations that have occurred in a limited number of cities, especially the so-called global cities or mega-agglomerations of North America, Europe, and the Pacific Rim (McCann, 2002). Researchers are beginning to look at the ways in which neoliberal urbanism develops unevenly, and part of this involves a return to matters of spatial politics and to the expression of urban political interests (Jonas & Ward, 2002).

Researchers are now discovering that not all cities are "entrepreneurial"—at least not in a narrow economic sense. Many so-called entrepreneurial cities are in fact engaged in various environmental and social improvements. These improvements are designed to make urban spaces more habitable for all residents (including non-humans) rather than competitive for the benefit of a few. Moreover, entrepreneurial cities are increasingly involved in attempts to reduce or restrict rather than to intensify their global impacts (e.g., impacts in terms of carbon emissions). There is growing interest in the uneven distributional consequences of neoliberal urbanism and in how different environmental and social interest groups across a city-region position themselves politically and strategically in relation to processes of urban economic transformation. There is also greater attention to the nature of the "social" and "environmental" in the contemporary politics of urban development, including a return to concepts of collective consumption and increasing interest in environmental sustainability (Cox & Jonas, 1993; While, Jonas, & Gibbs, 2004a, 2004b). Some researchers have begun to consider whether more democratic and socially inclusive spaces of governance are possible in the city and what role environmental and social issues play in this urban democratization process (Lake, 2003; Pincetl, 2003). Others have begun to talk in terms of urban governance "after entrepreneurialism," posing the question of why some cities have not aggressively pursued competitiveness agendas and instead pursued other (neo-Keynesian or arguably more redistributive) types of urban political strategies and interventions, including those of environmental sustainability (Haughton & While, 1999). Here it is felt that urban politics (e.g., strategies to secure social reproduction, environmental resources, and collective consumption in the city) do matter and that the difference such politics make are substantive rather than contingent.

In terms of rethinking the trajectory of urban governance after (or against) entrepreneurialism, one important idea is that of the sustainable city (Haughton & Hunter, 1994). This idea draws upon various principles of planning and urban management arising out of interpretations of the Brundtland Commission report, including its definition of sustainable development as meeting the material needs of present generations without compromising those of future generations. In this interpretation, the sustainable city should be designed in such a fashion as to minimize the global ecological footprint of urbanization (Rees, 1992). Instead of consuming resources and contributing to environmental degradation, cities ought to be managed in a way that enhances their role in sustainable development (Satterthwaite, 1997). As McManus (2004: 5) puts it, cities must be transformed from "growth vortexes" (i.e., places that consume resources without renewing or recycling them) to places that "*must* contribute to sustainability." That is to say, cities should contribute to development on a global scale that meets the daily needs of people without compromising the ecological, social, and economic needs of future generations.

In Europe and North America, urban governance for sustainability is often taken as meaning cities that have adopted formal political strategies on the environment. Examples include urban authorities signing up to international agreements on locally sustainable development, such as Local Agenda 21, or places participating in global climate change initiatives or annual World Urban Forum conferences. In addition, sustainability has emerged as an important agenda in urban planning, with increasing international interest in redensification, urban compactness, smart growth, green suburbs, healthy and sustainable communities, and the new urbanism. There is even some evidence that urban managers and political leaders in certain privileged "global cities" are devising socially and environmentally effective and inclusive planning, decision-making, and governance structures; that, in other words, there is much more going on in cities than a narrow economic reading of the literature on urban entrepreneurialism might allow.[2] Perhaps, after all, the new urban politics is as much about sustainability as it is about urban competition.

Tensions around Urban Entrepreneurialism and Sustainability

This is not to idealize the new urban politics of sustainability. Pronouncements to the effect that the age of urban entrepreneurialism has already given way to a new age of ecologically sensitive urban development are premature.[3] Even if in some cities attributes of the "right" urban governance for *environmental* sustainability can be found, they often coexist with neoliberal urban forms that are *socially* regressive and with which they may be in conflict. In practice, the urban taking up of strategies for sustainability such as Local Agenda 21 and climate change initiatives has been very patchy (Bulkeley & Betsill, 2002; Lake, 2000). Urban authorities no less than national governments are strategically selective in how they respond to state incentives on the environment (Jonas, While, & Gibbs, 2004). This suggests that knowledge of an urban politics constructed around issues outside the narrow focus on entrepreneurial economic development is crucial.

As with the literature on urban entrepreneurialism, there are reasons to be critical about the literature on urban governance for sustainability. Although academic attention to the environmental and social consequences of urban development is to be welcomed, some urban scholars and practitioners are guilty of conflating urban environmental sustainability with a concern for liveability and competitiveness. For example, in 2001 the Organization for Economic Cooperation and Development (OECD) produced guidance on how to move toward "better governance for more competitive cities" (Organization for Economic Cooperation and Development, 2001). According to the OECD, cities and metropolitan areas can be described as both competitive *and* livable if they include or embrace the following attributes:

- Economic diversity, especially in high value-added sectors;
- Skilled workforces, especially in knowledge industries;
- Strong institutional networks between research, industry, and education;

- The right environment for working and living, culturally, socially, and environmentally;
- Good communications and infrastructure;
- Institutional capacities sufficiently developed to deliver economic and social development over the long term. (OECD, 2001: 63)

The idea that all cities can be competitive, liveable *and* good for the environment seems to us to involve a leap of faith. Assuming that OECD guidance is taken seriously, it appears that urban managers are being steered toward at best a rather selective vision of urban sustainability in which "light-green" policies aimed at making cities more liveable can substitute for more radical actions on the environment that could by implication threaten the competitive approach to urban economic growth. There is also a danger of conflating sustainability with ecology and relying on social policies to address economic rather than environmental problems.

In many respects, the boundaries between the sustainable city and the entrepreneurial city have become—and for that matter probably have always been—blurred. These represent ideal typologies that may be useful for classifying different models of urban governance as "right" or "wrong" but do little in terms of investigating underlying political interests and rationalities. Insofar as they represent theoretical typologies rather than actual interests, processes, and mechanisms, they can serve only a limited function in terms of either explaining or investigating the actually existing spatial dynamics of politics in those Western cities undergoing rapid economic, social, and environmental transformations.

Given that we have these serious concerns about the ways in which concepts of urban entrepreneurialism and sustainability have been deployed in urban policy circles and in the literature on urban politics, a series of critical considerations follows on from this discussion:

- Is the pursuit of urban sustainability simply a legitimation strategy for cities, which are otherwise engaged in economic and cultural transformations designed to promote competitiveness? And, if so, to what extent is urban sustainability being mainstreamed or normalized as part of neoliberal urbanism?
- Under what conditions and using what kinds of resources can urban managers promote competitive *and* liveable cities? To what extent does liveability entail action around the urban environment outside the rules and constraints of neoliberalism?

- Are urban politicians responding to wider ecological, social, and economic developments primarily because of international and national political pressures, or are there genuinely local demands and pressures for sustainability to which they are responding? If the latter, where in the city do these demands come from?
- To what extent does environmentalism and the search for an "urban sustainability fix" (see While et al., 2004a) open up alternative pathways for cities? To what extent will these possibilities increase in the future?

Without attempting to answer these questions now, we turn to the case of Barcelona to illustrate how these questions also arise from a critical reading of contemporary conditions of urban entrepreneurialism.

LOOKING FOR SPACES OF SUSTAINABILITY POLITICS IN ENTREPRENEURIAL BARCELONA

Barcelona represents an interesting case study to examine in a chapter on critical approaches to urban sustainability. The city is increasingly written about as an exemplar of urban economic, social, and political transformation (Marshall, 2004) and an urban transformation model, moreover, that urban planners and managers in many other cities would like to replicate. Yet, it not necessarily viewed in the same terms as a model of urban environmental sustainability. In reality, urban authorities in Barcelona have paid increasing attention to the socioenvironmental consequences of urban economic transformation, as evidenced by the adoption of initiatives like Local Agenda 21, in attempts to engage with civic and environmental organizations throughout the city and wider region, and in the kinds of rhetoric espoused by local political leaders. In particular, there is increasing acknowledgment of problems of social exclusion, environmental damage, and political alienation caused by the city's dramatic transformation over the past 20 or so years. However, the municipal authorities have tended to address such problems through various planning and architectural "solutions" that have been imposed on people, neighborhoods, and districts throughout the city. While such interventions *have* incorporated principles of urban sustainability (e.g., into the design of buildings, infrastructure, parks, and open space), they are at the same time seen by critics of the city's bold transformation model to be part of a wider political and economic strategy to democratize urban space primarily if not solely for the purpose of valorizing the built environment.

For their part, local authorities have been slow to acknowledge that recent efforts to transform Barcelona economically and politically have not so much resolved as revealed growing social and environmental conflicts at specific locations both within the city and across the wider metropolitan region. In this respect, Barcelona's claim to be "La Ciutat Sostenible" (the sustainable city) is not itself a direct challenge to the existing model of urban economic transformation but instead is a response to the particular social and environmental conflicts produced by that transformation process.[4]

Entrepreneurial Barcelona: A Model of Urban Political and Economic Transformation

After the death of General Francisco Franco in 1975, democratic elections were held nationally and locally across Spain. In 1979, a populist socialist government was elected in Barcelona and ushered in a period of economic and cultural transformation in the city. With allies in international capitalist interests and the provincial government of Catalunya, the urban regime has looked for a postindustrial economic future for Barcelona. The transformation model followed on from, and in many ways was a reaction to, Franco's own attempts to integrate the city into the Spanish state's cultural, economic, and environmental modernization project. Throwing off its image as the city governed from and occupied by Madrid (Montalbán, 1987/2004: 161–197), the "new Barcelona" has evolved a different relationship to its regional and global environments.

A self-consciously international strategy of urban regeneration was pursued by a progressive, leftist, and Europeanist urban political regime led initially by Pasqual Maragall (mayor from 1982 to 1997 and now the president of the regional government in Catalunya) and his successor, Joan Clos (1997–2006).[5] A progressive element in the city's planning and architectural elite was revived and embarked on radical economic and urban transformations. Notably the city administration undertook major political and governance transformations, such as removing the centralized structures of metropolitan and regional planning put in place during the Franco era, abolishing the metropolitan corporation that had governed service provision across the wider urban region, and decentralizing urban administration to 10 districts across the central city. The latter strategy was an attempt to incorporate the local citizen movements that had flourished during the 1960s and 1970s as demands increased for services in the different neighborhoods around the city (Calavita & Ferrer, 2004). What planners and politicians now described as "the democratization of urban space" became a mechanism for radical transformations in the built form of the city and the restructuring of its economic base away from traditional manufacturing toward new cultural industries and property development. Transformation was partly about disrupting the rational metropolitan order underpinning the city's development under fascism (Montalbán, 1987/2004) but also was an attempt to reposition Barcelona as the economic and cultural capital of an imagined European urban axis stretching from London to Milan, and to make it the capital of the "north of the south of Europe" (Hughes, 1992: 36).

The city's bid to host the 1992 Olympic Games is widely credited as a catalyst for urban transformations. However, many activities led up to and followed on from this one event (see Table 5.1). These included the "opening up" (through demolition of older housing stock) of urban districts in the Ciutat Vella (i.e., the "old city," which is composed of residential areas until recently mainly populated by immigrants, the elderly, and the service-dependent), the construction of new hotels and "social" housing for young families and professionals, the elimination of remaining shanty housing on the Mediterranean waterfront, and the opening up of once polluted beaches to tourists and visitors. Major urban redevelopment projects were undertaken, which were financed by various international and Catalan business consortia working in partnership with regional and local government. These included the construction of a hotel–office–marina complex on the Front Maritim (Figure 5.1), the Olympic Village development, redevelopment of the old textile districts of Sants and Poble Nou around new-economy industries[6], and major new developments proximate to the mouth of the Besos River, which were constructed for the 2004 Cultural Forum. The city also embarked on transformations in the wider region, such as expansion of the airport and construction of a new high-speed international rail link.

Barcelona can justifiably be claimed to have undergone a political, economic, and social transformation: from a peripheral regional industrial center of Spain into a major cultural and economic hub of a new Mediterranean–northwest European urban axis stretching from Barcelona up to Paris and London and across to Milan and Munich. Transformation has capitalized on the city's unique architectural, industrial, and urban heritage as well as its fortuitous location next to the Mediterra-

TABLE 5.1. Major Moments in the Making of Barcelona into "La Ciutat Sostenible"

- 1979: Election of socialist urban administration led by Partit dels Socialistes de Catalunya (PSC) and Mayor Narcis Serra.
- 1982: Pasqual Maragall becomes mayor and declares Barcelona part of a new urban axis stretching from the Mediterranean to Northwest Europe.
- 1984–1986: City administration is decentralized and 10 neighborhood district associations created; city wins bid to host the 1992 Olympic Games.
- 1986–1992: Redevelopment projects leading up to the Olympic Games include Port Vell, Front Maritim, Olympic Village, renewal of Ciutat Vella, revival of Ildefons Cerdà's plan to extend Eixample and Diagonal Avenue to the sea, and redevelopment of industrial areas of Sants and Poble Nou.
- 1992: Olympic Games held in Barcelona.
- 1993: Plans under way to expand airport, create a new industrial logistics zone, clean up and divert the Llobregat River, and develop new wastewater treatment facilities.
- 1993–1996: Under increasing pressure from local activists and unions, the city eventually signs the Aalborg Charter and adopts Local Agenda 21.
- 1997: Pasqual Maragall is replaced by Joan Clos as mayor of Barcelona; Clos makes speeches on social inclusion.
- 1998: Exhibition *La Ciutat Sostenible*.
- 2000: Extension of Avenue Diagonal to the sea is completed, and clearance and redevelopment of Poble Nou and adjacent residential districts continues.
- 2000–2004: Completion of proposed new high-speed rail link to Barcelona from France and ongoing development of the 22@project in Poble Nou. City completes redevelopment of Besos River and new developments for Cultural Forum.
- 2005: Struggles around gentrification and renewal of Raval district in Ciutat Vella intensify, alongside growing criticism of city policies toward immigrants and public space; corruption around major public works projects exposed.
- 2006: Joan Clos ends term of office as mayor of Barcelona.

nean Sea. Yet, at the same time, there has been growing criticism about aspects of the social and environmental sustainability of urban transformation; criticism we now suggest to which the city has responded through attempting a marriage of urban entrepreneurialism and sustainability, albeit it is an uneasy relationship at best.

The Role of Sustainability in Urban Transformation

Toward the latter end of the 1990s, the political discourse around Barcelona's transformation made increasing reference to social and environmental problems by means of the trope "La Ciutat Sostenable" (The

FIGURE 5.1. Front Marítim in Barcelona including the developments associated with the 1992 Olympics and 2004 Cultural Forum (photo: © Andy Jonas 2006).

Sustainable City). Such attention to social and environmental sustainability became essential not just to the remaking of urban space inside Barcelona but also to the restructuring of metropolitan space to facilitate capital accumulation on a regional scale. Indeed, one important condition for the urban scaling of sustainability politics is that the economic influence of the city has spread out into the surrounding Catalan countryside. Definitions of the Barcelona metropolitan area have been revised over the years, but in general it has grown from 62 municipalities in 1981 to 216 in 1996. At the same time, the metropolitan population has expanded from 2.5 to 4.3 million residents (Nello, 2004: 32). This metropolitan growth provides a context for a growing sense of uneasiness about the city's relationship to its regional environment, not least because transformation has clearly benefited business interests locally dependent on Barcelona rather than on neighboring municipalities.

Pressure on the city to develop greener and more socially inclusive urban policies has tended to come from citizen and ecology groups within the city. In 1993, a popular citizens' coalition composed of ecology groups, the neighborhood associations, and trade unions put pres-

sure on the authorities in Barcelona to join the international campaign
Cities for Climate Protection (Tello, 2004). This campaign had been set
up after the 1992 Earth Summit and encouraged cities around the world
to implement measures to reduce global warming. In 1996, the city government signed on to the Aalborg Charter and defined a series of political actions including adoption of Local Agenda 21 to promote environmentally and social compatible urban development.

Since these early forays into the global sustainability agenda, urban
planners, politicians, museum directors, city managers, and businesses in
Barcelona have been eager to show off their city as an exemplar of good
practice relating to urban sustainability. In typical fashion, the city organized a public event to excite interest in sustainability. From April 1 to
September 13, 1998, the Centre de Cultura Contemporània de Barcelona held an exhibition titled La Ciutat Sostenible. The exhibition included a series of dramatic visual portraits and graphics demonstrating
how the city of Barcelona had transformed its physical environment and
natural setting over the course of the 20th century. Visitors to the city
were thus encouraged to participate, and be educated, in the new urban
spectacle of sustainability. This was not the first time the city had used
spectacle to reinvent itself, but this time it was reinventing itself as the
sustainable city.[7]

The exhibition posed the rhetorical question "Do we understand
what is happening to us?" and identified a number of environmental
problems organized around three spatial scales: local problems (e.g.,
dysfunctional transport, social segregation); regional problems (land
consumption, consumption of natural resources), and global problems
(the ecological footprint, climate change) (CCCB, 1998). These problems, the exhibition implied, were caused by the growth of the city. Tello
(2004) documents in some detail how environmental problems have indeed come to characterize the city's recent growth. He argues that "from
1996 onwards, the recovery witnessed in the economy has meant an increase in Barcelona's contribution to global warming" (2004: 234). He
further reports that greenhouse gas emissions in Barcelona increased
from 1.9 tonnes per inhabitant in 1987 to 2.88 tonnes per inhabitant in
1999 (Tello, 2004: 235, Table 13.1). Tello calculates that to sustain the
consumption habits of each resident of the city some 3.25 hectares person were required in 1996, making the city's ecological debt a total of
−1.53 hectares per capita (Tello, 2004: 240). The conclusion is that Barcelona has in fact become an *un*sustainable city.

What else besides public events has the city done to address this challenge of unsustainability? The city has engaged in Agenda 21 initiatives and networks across the region and Europe. Arguably, Barcelona has been at the forefront of the world's cities in signing international agreements on global sustainability. Indeed, since 1993 the strategy of urban transformation has increasingly incorporated—albeit selectively—discourses and to some extent also policies on social inclusion and sustainability. Principles of urban compactness, car-free streets, and social cohesion and inclusion have been applied to different planning solutions pursued throughout the city. Social inclusion and sustainability have likewise been important themes in the speeches made by Mayor Joan Clos, who has used international venues such as the United Nations World Urban Forum and the 2004 Cultural Forum (hosted by Barcelona) to emphasize the importance of cities as vehicles for social inclusion, good governance, and sustainability.

For the most part, it has been a combination of planning, architecture, and public spectacle that is the main tool for promoting sustainability in Barcelona. The city's planners and architects have in particular used neighborhood-scale urban interventions (rerouting traffic, planning for open space, redesign of land use, and creation and renovation of public parks) to stimulate city-wide economic transformations. Moreover, the city has identified so-called green areas for environmental improvement where nature parks, green corridors, and open spaces have been created or regenerated, sometimes making use of existing river corridors or recently demolished industrial and residential areas. Finally, public events such as the Forum of Cultures have become an opportunity for political leaders to declare publicly, on an international scale, progress toward a more sustainable urban future.

In some respects, Barcelona is typical of the challenges facing politicians and state managers in older urban industrial regions of Europe and North America. These have faced particular difficulties in terms of improving their economies through higher levels of environmental protection. This is due to the legacy of centuries of industrial contamination as well as negative images of the industrial city that are hard to overcome. Image is crucial. Barcelona's political and economic elite has struggled, and by and large succeeded, in overcoming its city's 19th-century legacy as the "Manchester of Catalunya."

Barcelona may not be a model of sustainability—yet, central to the urban regime's strategy is the mobilization of environmental assets for material goals. For example, the city's attempts to reclaim its maritime heritage and open up development to the sea make reference to the ancient geographical–economic connections across the Mediterranean, upon which the early prosperity of the Catalan state was forged. In this sense, the strategy of "returning the city to the sea" is a powerful environmental signifier, which underpins the new urban accumulation strategy. Likewise, attention to the regional environment signals a new relationship of the city to the provincial region. Reimagining Barcelona's natural environment in these ways has in effect meant that the city's political elite has mobilized the environment as a force of urban economic and political change, all the while appealing to Catalan sensibilities relating to the importance of nature, the sea, and mountains in the forging of national identity and, to some extent also, social and cultural cohesion.

But despite appealing to nationalistic values designed to unify, the transformation model incorporating urban sustainability does not go uncontested. Toward the end of his term of office, Mayor Clos drew increasing criticism from the Barcelona public, ecology groups, housing activists, immigrants, and many other organizations. These claimed that events such as the Cultural Forum were just another excuse to use public funds to attract visitors rather than attempts to transform urban space for the social, economic, and ecological benefit of local people (Qushair, 2006). In fact, the urban political landscape in Barcelona is defined by a number of ongoing struggles around social and environmental issues. We now describe two such struggles.

TENSIONS IN "LA CIUTAT SOSTENIBLE"

Social Struggle: Residential Displacement in Inner Barcelona

In an attempt to breathe new life into a socially stigmatized district in the heart of La Ciutat Vella, the city has carved out new public spaces in El Raval. This area has long been associated with criminality and prostitution and has been a cause of great concern on the part of those engaged in urban transformation, not least city authorities increasingly worried about what the planners call the "degradation of public space." A vital step toward the "clean-up" of the district was the construction of the Rambla del Raval. This was created by demolishing some of the existing housing stock in the late 1990s, opening up a space the size of about three street blocks (Figure 5.2). There is evidence that this project was designed to be more than another experiment in "the democratiza-

FIGURE 5.2. Rambla del Raval (photo: © Andy Jonas 2006).

tion of public space" within Barcelona. It is in fact a project steeped in the urban instruments of social control. For, as if serving as a dramatic if not in fact conscious reminder to the people, the creation of Rambla del Raval has opened up a new vista toward the church located on the top of Tibidabo (a hill rising above the city). This church was built and funded by the city's religious and conservative elite in atonement for the sins of the city's working class following the urban protests held during Tragic Week in 1909.[8] It serves as a reminder to the residents of Raval of the consequences of urban unrest.

In the "new" Raval, a five-star hotel has been constructed, and young professionals and tourists are encouraged to move into an area otherwise populated by longtime residents from the different provincial regions of Spain but increasingly popular for recent immigrants from Africa and South Asia, many looking to eke out a living from the tourist traffic. Local housing activists, anarchists, social planners, and groups representing immigrants have described forced displacements of local residents by the city authorities and local landlords.[9] Rental and apartment prices have been driven up by speculation and the relentless force of international property capital. In many respects, Raval has become a battleground for the future of a more socially sustainable Barcelona.

The city has relied on the neighborhood associations to channel urban protest in directions that it sees as more productive and engaged with transformation. Since these organizations have, for all intents and purposes, been formally coopted, residents, activists, and protesters in Raval have defined their own spaces of politics and struggle. In the Spring of 2005, local streets were barricaded by protesters, thereby disrupting local tourist traffic and construction work. Protests were also directed at the city government. According to Qushair (2006), from June onward, grassroots organizations staged protests every Wednesday calling for Mayor Clos's resignation. The protests drew up to 2,000 participants and challenged the city's policies on drug rehabilitation and freedom of speech as well as the aggressive tactics of the police toward immigrants.

Years of struggle more generally against the transformation of Barcelona reached an apogee early in 2005 when a large social housing complex in the Carmel district of city showed signs of structural weakness, forcing some 1,200 residents to seek alternative temporary accom-

FIGURE 5.3. Parc del Llobregat. (photo: © Andy Jonas 2006).

modations. The structural failure was attributed to a subterranean collapse resulting from the extension of a metro line near the complex to serve new urban developments in the northeastern inner suburbs. There was a flurry of articles in the newspapers, which accused Catalan state officials of bribery and corruption over the granting of contracts for European Union-funded public works projects. Mayor Clos was also a major target for criticism (e.g., over plans to increase parking charges in the old city), eventually leaving office in 2006.

Ecological Struggle: Airport Expansion and Suburban Nature

As we have suggested, the city of Barcelona has sought to introduce formal governance processes such as Local Agenda 21 so as to circumnavigate a variety of potential ecological and social conflicts within its jurisdiction. However, there is another story to tell about the suburban areas under pressure from the city's aggressive economic development policies. One of the largest and most costly economic development projects has involved redevelopment of the airport and surrounding industrial zones, including Zona Franca and the port. After the recent completion of an extra runway and terminal building, Barcelona's airport is now one of the largest and busiest in Europe in terms of passenger numbers,[10] and the city has benefited economically from the expansion of low-cost airline services. The growth in visitors and conference activities has likewise boosted the hotel and hospitality industry in the city, making Barcelona a popular location for the development and expansion of national and international hotel chains catering to a range of customers.

Barcelona airport is located near the coastal zone in the Llobregat River delta, an ecologically sensitive area that attracts birds and other wildlife (Figure 5.3). In the lead-up to the Olympics in 1992, there were plans to realign the Llobregat River to make way for airport expansion (see Marshall, 1993), but this brought the city into conflict with neighboring local authorities such as the community of El Prat, politically in the control of the local Communist party. The conflict was about competing uses of areas under threat from the expansion of the airport, areas used by residents for recreation, local restaurants and other business, and so forth. There was also a debate over the development of a new water treatment facility.[11]

Authorities in Barcelona responded to the conflict by trying to represent the airport development as a "win–win–win" solution, or, as one brochure put it, "Transformarse para ganar" (to win from transformation). In an attempt to appease ecologists and local residents, a nature park was established in close proximity to the airport. As if in a subversive vein, the main road alongside the airport, which has access to the nature park, is a popular location for prostitutes and pimps, albeit we can only speculate whether these are the displaced of El Raval.

What can we read into these sorts of tensions and struggles in Barcelona? Are they a sign of the failure of urban political leaders to address the social and environmental problems created by displacements in the old city and ecological problems in outlying urban areas related to suburban growth, new transport links, and expansion of the airport? Are they, in this respect, a sign of politically *un*sustainable urban transformation? To what extent has the city selectively used the discourse of sustainability to legitimate economic transformations even as transformation has generated growing conflict on a metropolitan scale?

There can be no doubt that ecological and social demands *are* creating new pressures on entrepreneurial urban governance in places such as Barcelona. But should we see these demands as relatively autonomous from the economic pressures that make cities entrepreneurial? Should we develop different concepts with which to examine environmental and social sustainability and separate these from knowledge of urban entrepreneurialism? What exactly is the relationship between the "sustainable city" and the "entrepreneurial city"? And if these two ideal types are no longer useful for understanding the contemporary dynamics of urban politics, what should replace them? In an attempt to answer some of these questions, we now resort to more abstract considerations.

RETHINKING THE CITY AS A SPACE FOR SUSTAINABILITY POLITICS

Upon reflection, critical urban theorists are justified in appearing skeptical about the need to incorporate social, economic, and environmental sustainability into their conceptual frameworks of urban governance. In fact, most of the so-called definitive statements about after-Fordist urban governance have had very little to say about the natural environment or indeed, for that matter, sustainability (these two concepts should not necessarily be seen as commensurable).[12] They have instead focused on the economic (accumulation) and to a lesser extent the social (collective consumption) contradictions found in entrepreneurial cities. For these scholars, the environment and its politics are rarely examined, other than perhaps as contingent outcomes of economic change in such cities, as extraeconomic conditions rather than constituent causes of urban development and its politics.

Evidently, the existence of an actually existing urban politics of sustainability would present a number of challenges to critical urban theorists.[13] Perhaps this should not surprise us. There is a long tradition of critical writing about the Western city that has excluded knowledge of nature and the environment. This is especially true of classic treatises on the conditions of the working class in the 19th century by Marx and Engels. It also applies to formative works on the state by the likes of Weber and Gramsci. In the latter part of the 20th century, environmental historians began to recover some sort of critical-intellectual perspective on the role of nature in the making and transforming of Western cities (e.g., Davis, 1999; Worster, 1985, 1994). These ideas increasingly have informed the development of a growing body of work in geography around urban political ecology, a proper discussion of which is beyond the scope of this chapter.[14]

Not that critical work on urban growth has ignored the environment altogether. For example, work on the "city as a growth machine" in North America commented on the rise of environmental interests and coalitions during the 1970s and 1980s (Molotch, 1976). These potentially represented a serious challenge to those of rentiers, that is, the promoters of urban growth, including developers, city governments, politicians, etc. (Logan & Molotch, 1987). Yet, the environmental "countercoalition" was not able to reverse the trajectory of urban growth significantly, in part because growth coalitions could harness nonlocal interests in support of their growth strategies (the state and federal governments, multinational firms, etc.) or simply relocated their profit-making activities elsewhere and spread investments across a wider territorial arena.[15] However, abstracting such "environmental" interests and their demands from the totality of the urban political experience continues to be problematic for many scholars of the Western city. Much work remains to be done on specifying how tensions between ecological, economic, and social demands are expressed concretely in and through spatial interests, political practices, and territorial coalitions across the entrepreneurial city.

Part of the problem in thinking about the relationship between urban political interests and spaces of sustainability politics is that much of the work on the new urban politics has tended to locate (in a theoretical sense) "environmental" issues in a politics of the living space, which is in turn separated analytically from the politics of production (Carlin & Emel, 1992).[16] Thus, various environmental improvements are seen to alter or affect the consumption habits and lifestyles of urban and suburban residents, perhaps indirectly profiting them by boosting home property values or lowering the costs of household services, and thereby also improving the city's liveability and attractiveness in terms of economic development (Heynen, 2006). These are factors that obviously can play into the hands of the promoters of urban and suburban development—the growth coalitions or rentiers—who appropriate exchange value accordingly (Logan & Molotch, 1987).

By the same token, negative environmental externalities such as pollution, congestion, and waste are problems to be banished from the city undergoing redevelopment and integration into the new economy, especially where they are in proximity to areas either already or soon to be populated by educated and environmentally aware consumers. This way of thinking about the environment as an externality, and one which can be mobilized and manipulated for the benefit of promoters of economic development, is mirrored in formal strategies and urban policy interventions such as the smart-growth movement, ideas of the compact city, the greening of suburban space, urban river restoration, or attempts to covert old industrial uses into modern living spaces.[17]

Furthermore, these ways of thinking about the environment (as, i.e., in an oppositional relation to urban competitiveness) have only served to reinforce an ontological separation of the "urban" from the "environmental," with implications for the extent to which "environmental" conflicts are in fact no longer seen to be related in any way to deeper racial and class divisions and conflicts arising from production and economic development (or the lack thereof) in the city[18] (Lake, 2003). Moreover, the lack of attention to how environmental issues and conflicts in the city or wider region are socially constituted through state strategy and collective action has been a significant weakness in the literature (Gibbs & Jonas, 2000).

Overall, there is a distinct danger that urban theorists, if they consider a politics of sustainability as such, see it as one fought exclusively around "environmental" issues devoid of any social (and spatial) context. The danger here is that modern environmental improvements, such as the creation of green open space and parks or attempts to set up and attract clean or ecologically friendly industries, otherwise seen as beneficial to the economic and environmental health of a city, can in some circumstances still operate as strategies of social control rather than as means of political empowerment. Such strategies—perhaps conducted in the name of greening the (entrepreneurial) city—might in effect discipline the poor, the elderly, ethnic minorities, and the socially marginalized and displace or exclude them from areas of the city undergoing so-called environmental improvements.[19]

Nevertheless, an urban-scaled politics of sustainability fought around social *as well as* environmental issues can offer opportunities for developing counterstrategies on behalf of marginalized groups (Pincetl, 2003). In this vein, Raco (2005) has usefully highlighted some potential contradictions and possibilities arising from the state's attempts to steer urban growth along a pathway of sustainable development (SD). He comes to the conclusion that

> the extent to which SD agendas and frameworks take on neoliberal forms becomes an empirical question to be interrogated in and through specific case studies. For instance, if neoliberalism does shape and dominate the policy activities of Western governments, we might expect that SD will be deployed and reinterpreted in ways that challenge the legitimacy of state regulation and control, and promote market-driven development agendas. . . .
>
> Yet, at the same time the emergence of SD, at a variety of different scales, provides policymakers and a range of communities with alternative ways of thinking about economic development, social justice and resource use. The enhanced focus on the impacts and externalities generated by economic development can challenge neoliberal inspired growth agendas and modes of regulation. (Raco, 2005: 330)

In summary, if there is a general tendency toward a convergence of sustainability and competitiveness around the city, this is not happening on the ground in a straightforward, unproblematic, and uncontested way. Many specific issues remain to be fought out at and around the urban scale. It is in this context that we discuss three further examples of urban-scaled sustainability politics with a view to identifying a variety of empirical possibilities.

NEW SPACES OF SUSTAINABILITY POLITICS IN THE CITY

The first example considers the ways in which new state strategies toward sustainable urban growth are bringing about new strategic politi-

cal alignments and opening up the city-region as a particular territorial
scale around which important issues such as collective provision are
fought. The second example considers what the new politics of mobility
(i.e., transport policy) in the city means from the standpoint of understanding strategic shifts in the composition and strategies of business-led
growth coalitions. Finally, we examine the circumstances under which
ideas of "just" sustainability have entered into mainstream urban governance and what tensions have resulted.

Sustainable Development and the Footprint
of Urban Governance: The Politics of Collective
Provision in England's South East

The argument of this chapter so far has been that tensions around competitiveness and sustainability have become focused at and around the
urban scale, more or less defined in terms of economic growth. Now
some scholars believe that competition and economic growth in the new
global capitalist order are driven by the growth of megaurban agglomerations or city-regions (Scott & Storper, 2003). Local authorities in
towns, suburbs, cities, and rural areas within or proximate to mega-
agglomerations are just as likely as the larger city governments nearby to
be interested in developing entrepreneurial approaches to economic development, collective provision, and ecological protection. In these circumstances, it is perhaps more appropriate to talk of *entrepreneurial
city-regions* (see Jessop, 1997b), and hence the possibility that sustainability is fought around new spaces of politics and state that more or less
coalesce at the city-region scale. In this respect, what constitutes the "urban" as a separate scalar category or abstraction is itself debatable.
However, there is certainly increasing evidence that the demands of economic development and environmental protection, and moreover the
connections between the two, are leading to experimentation with new
spaces of governance that blur the distinction between the "city" and its
"region" or hinterland.

In the United Kingdom, the rolling out of neo-Keynesian policies for
sustainable development on the part of the state to address various failures of neoliberal urbanization has exposed various tensions, not least of
which are the growing social costs of reproduction of labor power functional to the success of new economic spaces in and around metropolitan
London and the South East region. As Raco (2005) shows, many com
munities in and around this large and growing urban region are confronted
with such tensions, including lack of access to quality public services,

problems of housing unaffordability, and lack of investment in basic infrastructure.

These tensions are especially acute in growth centers identified by the U.K. state as of national importance such as Cambridge, a city where pressures of economic development have produced new demands for housing and infrastructure. To some extent, these demands have come not so much from local communities (some of which are in fact opposed to growth), or even from organized labor and consumers of services, as from the growth interests themselves. We have suggested elsewhere that a new space of sustainability politics has opened up around struggles for funding collective provision: of providing solutions to a deficit in local and central state spending on infrastructure to service new housing and industrial developments in the growth corridors connecting Cambridge to other local centers of production in the wider region (see While et al., 2004b). In this context, new imaginaries of the city-region have legitimated neo-Keynesian institutional developments that are designed to address structural problems of service underprovision identified by economic growth interests.

A more general theoretical point can be made here that, by identifying a politics of sustainability in the entrepreneurial city, there is an opportunity for theorists to reconnect knowledge of the politics of urban economic development with that of collective consumption and environmental provision. Work on the neoliberal city suggests that the state is offloading its responsibilities for social reproduction onto new privatized arrangements and that city-regions are best understood as new sites of global accumulation rather than places where matters of social reproduction are fought out. However, we can suggest that in fact new structural contradictions have emerged around the environment and collective provision that neither the state nor capitalist interests dependent on city-regions appear able to resolve on their own. The result is that we might expect a proliferation of new spaces of politics and state across entrepreneurial city-regions around matters of social reproduction, with sustainability politics enabling the emergence of new and unexpected neo-Keynesian cross-class alliances and territorial coalitions (see Krueger & Savage, 2007).

Urban Politics of Mobility: Business Coalition Realignment, Sustainability, and the New Urbanism in the United States

One of the spatial contradictions of the entrepreneurial capitalist city is the necessity to invest in fixed infrastructure in order to foster mobility:

the circulation of commodities, access to raw materials, the journey to work, and so forth (Harvey, 1982). On the one hand, there are huge economic incentives underpinning the construction of freeways, rail links, and airports that connect the city to new suburban development, itself an important outlet for effective demand. On the other hand, exchange values may be appropriated by capitalists who depend on proximity to alternative and local transport links, including those in downtown areas. Periodically cities experiencing rapid growth can face a "crisis of mobility" (Henderson, 2004) as congestion increases and there is a lack of investment in alternative transport modes. In these circumstances, business interests dependent on the spatial organization of mobility in the city might intervene.

For proponents of the sustainable city, one might anticipate that transport policy would be a key focus of political strategy development. Indeed, typical indicators of *un*sustainable urban development often single out transport-related problems such as congestion, pollution, and suburban sprawl, which can in turn contribute to such health issues as stress and obesity, and hence also to a sense of environmental *in*justice. But as Jason Henderson (2004) points out, the politics of mobility are complicated, and there are all sorts of possibilities for unusual alignments to occur among different factions of capital. He refers in particular to the recent situation in metropolitan Atlanta in the United States.

Between 1990 and 2000, Atlanta was the second fastest-growing metropolitan area in the United States, and smog, congestion, and sprawl were major concerns, not least because of the threat of the withdrawal of federal funding. Facing a crisis of mobility, corporate business interests began to lobby for more compact and concentrated urban forms, including improvements in the inner-urban rail transit system, provision for cycling, and various other smart-growth policies. As Henderson (2004) argues, this brought about an unusual alliance between downtown business interests and environmental, civil rights, and neighborhood organizations in the city. However, there were other interests that favored more compact and auto-centric developments on the metropolitan Atlanta periphery. These lobbied for further expansion of the suburban freeway system and for bonds levied for extension of peripheral road networks. Tensions among factions of capital became focused on developments known as the northern Arc, which would have diverted funds away from a more compact urban strategy. Henderson concludes:

> The northern Arc debate shows us that divisions over mobility exist between factions of capitalist elites. . . . [I]t was more than a struggle of

suburban versus urban interests, but rather a struggle of competing conceptualizations of how space should be organised and what types of mobility space would be organised. . . . The motivation of capitalist interests in Atlanta is to overcome the contradictions and tensions between hard and soft mobilities and control implementation of mobility strategies that make the metropolitan region globally competitive and thus securing increased exchange value of the region. (2004: 208)[20]

The Search for Socially Just Sustainability in the Entrepreneurial City: Food Policy in Canada

If the Atlanta story exemplifies tensions among economic factions in the entrepreneurial city, one of the key issues for critical research on urban sustainability politics is identifying precisely where and when concepts and practices of environmental and social justice enter into mainstream politics. The principle of "just" sustainability draws attention to this: the importance of uniting a concern for environmental security with that of social equity (Agyeman, Bullard, & Evans, 2003). Here questions of access to the basic daily urban needs of food, nutrition, health, and security are paramount, especially for many people who live under conditions of neoliberal urbanism, wherein they are often deprived of such basics (often this is related to neoliberal policies such as the withdrawal of public services).

Wendy Mendes (2005) combined research and planning practice in Vancouver, Canada, to develop a fresh perspective on how food policy has become a vehicle for promoting social sustainability in the entrepreneurial city. Political and economic elites in Vancouver have engaged in entrepreneurial policies in the context of fiscal constraints imposed by neoliberal provincial and federal governments. Nonetheless, the city of Vancouver formally adopted a food policy in December 2003 when the city council voted to adopt a proposed "Action Plan for Creating a Just and Sustainable Food System for the City of Vancouver." This action plan had been prepared by social planners upon consultation with a Food Policy Taskforce made up of more than 70 food-related interests, including representatives from such organizations as farmers' markets, community kitchens, food banks, food processors, organic and sustainable food businesses, dieticians, health professionals, schools, environmental groups, and antipoverty groups (Mendes, 2005: 2).

Although the plan was supported by a broad coalition of progressive interests and adopted by the city council, a number of tensions soon became apparent. There were concerns expressed at council hearings about the perceived appropriateness of such a policy in a city committed to competitiveness and perhaps unable to assign resources to an activity not seen to be part of mainstream urban governance. The city had in particular suffered a downgrade in its credit rating, and there were concerns about what kind of message was being sent out to investors. Given the global implications of food production and consumption, why should an entrepreneurial city like Vancouver adopt food policy as a function of municipal government? In addition to concerns about competitiveness, a second important tension was an inability to reconcile the goal of sustainability with that of addressing hunger ("food security"). One of the reasons that food policy was ultimately accepted as a municipal function in Vancouver was because it could be aligned with preexisting policy directions and organizational expertise in sustainable development and not because it was understood as a tool to address social justice concerns like hunger and food insecurity. Where the latter were concerned, the municipality saw limitations in its capacity (or political appetite) to intervene.

Mendes (2005) thus shows how various tensions and contradictions are revealed as food policy (as an instrument of social sustainability) becomes scaled at the municipal level. As economic and political elites in entrepreneurial cities seek to be seen as good stewards of their environment, the urban scaling of sustainability politics becomes a necessary possibility; but the form that this scaling takes is contingent. In the Vancouver case, it provided opportunities for progressive food interests across the city to develop a strategy around notions of social planning for sustainability, linking the spaces of neighborhood and local consumption to wider spaces of the city and regional governments.

CONCLUSION

It is hoped that this chapter has raised some relevant issues about how critical urban theorists might want to approach the often contradictory relationship between urban entrepreneurialism and sustainability. The examples of changing spatial politics of sustainability in the city presented above serve to highlight a variety of possible tendencies that could be found in the governance and politics of the entrepreneurial city. Yet, when viewed in terms of the *longue dureé* of social and political developments under capitalism, there is nothing necessary or essential about seeing sustainability through the lens of the contemporary entrepreneurial city. Just as the relationship between industrial capitalism and the growth of the city was once contingent but has became necessary, so also is there no basis for assuming that sustainability politics is necessarily a politics of the city—*of* the urban as opposed to *in* the urban. That it might have become so as a result of the rise of urban entrepreneurialism in the West is quite likely. At some point in the recent history of the development of capitalism, a range of ecological, social, and ecological interests and demands have come to coalesce around and through the institutions and spaces of the entrepreneurial city. Accordingly, the relationship between urban development politics and sustainability has, we believe, become less contingent and more necessary to the realization of interests dependent on working and living in such places.

It is perhaps worth reflecting that toward the end of the 19th century discussions of the "environment" began to refer increasingly to the social problems of the industrial city—overcrowding, disease, poverty, social unrest, class struggle, problems of habituating workers to the discipline of factory production, and the like (Boyer, 1978). Industrialists and middle-class reformers became political champions of the idea of the socially functional and efficient city: improvements in water supply and sewerage disposal, new modes of transport such as metro systems and urban parkways, and the creation of green spaces, gardens, and the "city beautiful" took hold (Mumford, 1961). At this time, there was no sustainability discourse, but nonetheless attempts to achieve a harmonious balance between the city and nature through attempts to urbanize the latter were central to solving problems of social control. This is not to romanticize about 19th-century environmental reforms but rather to suggest that intervening years saw environmental issues largely banished from our knowledge of the city, including from our understanding of the rise of urban entrepreneurialism.

What has happened lately has been something of a recoupling of urban political reforms and environmental interventions. How this has happened reflects a complex conjuncture of interrelated processes and conditions. First, there have been increased concerns about global environmental degradation, which have meant that urban managers now face more stringent demands in terms of environmental protection and promoting sustainable urban development. These require a coherent and holistic approach to an urban system and potentially to a rethinking of state–citizen relationships at the urban scale as responsibilities for international environmental obligations are passed downward. Second, the city itself is again seen as a site of production, innovation, and accumulation. So, an emphasis, for instance, on carbon-reduction strategies means that urban "environmental" policy is likely to have potential implications for all areas of urban economic strategy (given the diversity of urban economic activities and uses dependent on fossil fuels). Third, the city is also being reinvented as a place of social mix, citizenship, diversity, and human creativity. Thinking about environmentalism in this context involves far more than examining the responses of upper- and middle-class residents in the suburban living place. Finally, urban residents of *all* classes and ethnic backgrounds have higher expectations with respect to the quality of life, and the economy (at the moment, capitalism) has the capacity to deliver on such expectations even if there is no political will for it to do so in a more equitable way.[21]

All of this means that in the realm of urban political space the (social) economy and the environment are being reconnected in various ways, with the glue being provided by discourses and practices of sustainability. Potentially this could have far-reaching implications for our understanding of the dynamics of urban politics, not least in terms of its formative spatial preconditions. For it is not simply a matter of whether or not urban governance is oriented more toward the ideal of the "sustainable city" than to that of the "entrepreneurial city" (as past theoretical categories would dictate), but rather instead that new spaces of politics for sustainability are opening up around new strategic territorial and class alliances and divisions. Hopefully, this chapter has helped in identifying where in the entrepreneurial city researchers can begin to look for such spaces. An important task for future research of a more critical orientation will be to identify whether or not the contradictions of urban entrepreneurialism are opening up possibilities for an urban-scaled politics constructed around a stronger and more socially just "sustainability fix."

ACKNOWLEDGMENTS

We extend special thanks to Bob Lake, Wendy Mendes, and Stephanie Pincetl for detailed comments on a draft of this chapter. Further comments were provided by Donald McNeill, David Gibbs, and Rob Krueger. Jan Merrigan introduced Andy Jonas to activists in Barcelona and helped with translation from Spanish and Catalan languages. A primitive version of the argument was presented at the Workshop on Metropolis to Exurbia: Sustainability of the Modern Urban Complex, Green College, University of British Columbia, March 2004. A more refined version was presented to the Centre for Urban and Regional Studies and the Department of Geography, University of Newcastle, New South Wales, in July 2006. We would like to acknowledge funding from the Economic and Social Research Council for the project "Governance and Regulation in Local Environmental Policymaking" (Award No. R000237997).

NOTES

1. From the ecological side, one of the key pressures for change is the impact of global and national carbon reduction strategies on urban systems historically built around spatial forms and land uses that rely on the burning of fossil fuels. While low carbon urban economies may be possible, the current transition phase requires some radical rethinking of the resource basis to urban economic development.

2. This is not to say that neoliberal policies do not impose resource constraints on cities engaged in environmental and social improvements. Moreover, urban environmental crises (e.g., Hurricane Katrina in New Orleans) can expose contradictions in neoliberal urbanism, allowing urban residents to criticize and challenge urban political leaders and state policymakers.

3. Our research in England, for example, has found evidence of a highly selective "sustainability fix" in which entrepreneurial urban regimes adopt—or coopt— nonchallenging environmental discourses and initiatives but shy away from deeper green actions that are perceived to impact on urban competitiveness (While et al., 2004a).

4. The ensuing discussion of Barcelona's urban transformation draws on various sources, including Montalbán (1987/2004), Hughes (1992), Marshall (1992, 1993, 2004) and McNeill (1999), as well as field notes, observations, and interviews compiled and conducted on regular study visits by Andy Jonas between 1998 and 2006.

5. It should be noted that the city's strategy of international if not global economic and cultural integration has carefully been nurtured around Catalan nationalist sensibilities, albeit that the marriage of nationalism, republicanism, and socialism in the city has not always been easy, often prone to disruption in moments of national and international crisis. In this respect, the new Barcelona remains as much a product of its unique historical, political, and cultural conditions as it is emblematic of entrepreneurialism writ large on a global stage.

6. The redevelopment of Poble Nou is known as the 22@ project, as it is based on an old planning area or zone (zone 22) that was recognized in the first metropolitan plan for Barcelona. The 22@ project seeks to integrate new-economy industries into the existing built environment of an old mixed industrial–commercial– residential district located adjacent to the Olympic Village redevelopment.

7. Other major urban propaganda events held in Barcelona have included international expositions in 1888 and 1929, the 1992 Olympic Games, and most recently the 2004 Cultural Forum. The exhibit on urban sustainability was subsequently relocated to a municipal building on the Ramblas, which is the main tourist thoroughfare in Barcelona.

8. The Tragic Week of July 1909 began as a protest against the Spanish War with Morocco but turned into a workers' general strike. Several buildings and convents were destroyed in the ensuing battle between police and protesters.

9. Personal interview with Raval activists in March, 2006. The ensuing narrative is based on field work conducted in 2004, 2005, and 2006.

10. In 2005, it was the ninth-busiest airport in Europe, with 27 million passengers per year, and it has the capacity to expand further to between 40 and 55 million passengers (*en.wikipedia.org/wiki/Busiest_airports_in_Europe*, accessed in October 2006).

11. Local residents interviewed by Andy Jonas in the vicinity of El Prat in 2004 claimed that their traditional access to the sea has been restricted by the economic development of the airport area and that local family activities (walking, fishing, picnics, etc.) and some local businesses (restaurants, campgrounds) have suffered as a result.

12. Some theoretical treatments of post-Fordist urban governance can be found in Harvey (1989), Mayer (1994), and Jessop (1997a, 1997b). For a prescient discussion of how the environment "fits" into post-Fordist urban governance, see Keil and Graham (1998).

13. This is not a problem exclusive to urban theory but rather applies also to subdisciplines such as economic geography. See Bridge and Jonas (2002) and Zimmerer (2000).

14. Drawing on neo-Marxist political ecology (see O'Connor, 1996), urban political ecology considers the necessary dependence of cities and their inhabitants on the production and consumption of nature at different phases of capitalist development (Harvey, 1996), including recent times (Gandy, 2002; Keil & Graham, 1998). For these scholars, the city forms a kind of "socio-natural hybrid" (Swyngedouw, 1996) wherein ecological demands can exert an autonomous influence on urban institutions, politics, and governance. Excellent examples of this literature can be found in Braun and Castree (1998), Swyngedouw and Heynen (2003), and Keil (2005).

15. These days one might argue that local growth interests have engaged in a politics of rescaling (powers and functions in) the state (Cox & Jonas, 1993). This chapter does not address state rescaling and its relationship to sustainability politics at the urban scale. Much of the recent work on neoliberal urban governance has examinee the process of state rescaling on an abstract level. In our view, the question as to whether the rescaling of the state necessarily results in an urban-scaled politics of sustainability is one that cannot be answered in the absence of knowledge of concrete spatial interests. Nor can it be considered without some attention to what we have referred to elsewhere as the process of "eco-state restructuring," wherein the state steers the local state and urban policy around more ecologically sustainable pathways (While et al., 2004a; Jonas et al., 2004; see also Meadowcroft, 2005).

16. This tendency is, to some extent, reproduced in work on urban sustainability politics, which concentrates on the consumption of urban environment at the expense of the knowledge of how, in the making of the entrepreneurial city, nature is produced.

17. A very readable and up-to-date review of the contribution of watershed planning to urban sustainability in the United States can be found in Platt (2006).

18. In the United States, responses to "environmental" conflicts in suburban areas

have been shown to differ among residents depending on race, class, and housing tenure (Feldman & Jonas, 2000; Neiman & Loveridge, 1981).

19. These issues have been thrown into sharp relief by Hurricane Katrina and its long-term impact on the built environment and people of New Orleans. Here questions have been raised about the social and economic implications of building cities in areas vulnerable to short-term and extreme natural events as well as longer-term processes of climate change. These implications, as Hurricane Katrina revealed, are not simply about conflicting scientific views and the policy impact of global environmental trends but are also about how these views are infused into the deeply racialized and class politics of spaces of the entrepreneurial city.

20. The costs of road congestion mean that mobility is one area where one might expect to see a convergence between entrepreneurial and sustainable city strategies. In London as well as other U.K. cities, road—or congestion—charging is increasingly being framed as a "win–win–win" strategy for urban leaders, ensuring easy mobility by road for those who can afford to pay for premium urban connections (see Graham, 2004) while generating tax income, reducing the carbon deficit, and bringing quality-of-life benefits to central cities. The acceptance of road charging by economic interests lies in the costs imposed by congestion on firms, commuters, and visitors. Crucially, however, it is also justified by discourses related to global environmental protection.

21. There is an indication that recent trends open up opportunities to address social and health problems in cities that could relate to the environment but that are also being examined in their own right. These social issues—perhaps even more than ecological issues—will be crucial in defining the urban sustainability politics of the future. These trends offer a significant challenge to earlier thinking on what constitutes appropriate municipal policies and development trajectories beyond the environmental but still in support of "sustainability." Thanks to Wendy Mendes for pointing this out.

REFERENCES

Agyeman, J., Bullard, R., & Evans, B. (2003). *Just sustainabilities: Development in an unequal world*. Cambridge, MA: MIT Press.

Boyer, P. S. (1978). *Urban masses and moral order in America, 1820–1920*. Cambridge, MA: Harvard University Press.

Braun, B., & Castree, N. (Eds.). (1998). *Remaking reality: Nature at the turn of the millennium*. London: Routledge.

Brenner, N., & Theodore, N. (2002). Cities and the geographies of "actually existing" neoliberalism. *Antipode, 34*, 349–379.

Bridge, G., & Jonas, A. E. G. (2002). Governing nature: The reregulation of resource access, production and consumption. *Environment and Planning A, 34*, 759–766.

Bulkeley, H., & Betsill, M. (2002). *Cities and climate change*. London: Spon Press.

Calavita, N., & Ferrer, F. (2004). Behind Barcelona's success story—citizen movements and planners' power. In T. Marshall (Ed.), *Transforming Barcelona* (pp. 47–64). London: Routledge.

Centre de Cultura Contemporània de Barcelona (CCCB). (1998). *La ciutat sostenible.* Barcelona: Author.

Cox, K. R. (1993). The local and the global in the new urban politics: A critical view. *Environment and Planning D: Society and Space, 11,* 433–448.

Cox, K. R., & Jonas, A. E. G. (1993). Urban development, collective consumption and the politics of metropolitan fragmentation. *Political Geography, 12,* 8–37.

Cox, K. R., & Mair, A. J. (1988). Locality and community in the politics of local economic development. *Annals of the Association of American Geographers, 78,* 307–325.

Davis, M. (1999). *Ecology of fear: Los Angeles and the imagination of disaster.* London: Picador.

Dryzek, J. (1997). *The politics of the earth: Environmental discourses.* Oxford, UK: Oxford University Press.

Feldman, T., & Jonas, A. E. G. (2000). Sage scrub rebellion?: Property rights, political fragmentation, and conservation planning in Southern California under the federal Endangered Species Act. *Annals of the Association of American Geographers, 90,* 256–292.

Gandy, M. (2002). *Concrete and clay: Reworking nature in New York City.* Cambridge, MA: MIT Press.

Gibbs, D. C., & Jonas, A. E. G. (2000). Governance and regulation in local environmental policy: The utility of a regime approach. *Geoforum, 31,* 299–313.

Graham, S. (2004). Flowcity: Networked mobilities and the contemporary metropolis. In T. Nielsen, N. Albertsen, & P. Hemmersam (Eds.), *Urban mutations: Periodization, scale, mobility.* Aarhus, Denmark: Arkitektskolens Forlag.

Hall, T., & Hubbard, P. (Eds.). (1998). *The entrepreneurial city.* Chichester, UK: Wiley.

Harvey, D. W. (1982). *The limits to capital.* Oxford, UK: Blackwell.

Harvey, D. W. (1989). From managerialism to entrepreneurialism: The transformation of urban politics in late capitalism. *Geografiska Annaler, 71,* 3–18.

Harvey, D. W. (1996). *Justice, nature and the geography of difference.* Oxford, UK: Blackwell.

Haughton, G., & Hunter, C. (1994). *Sustainable cities.* London: Jessica Kingsely.

Haughton, G., & While, A. (1999). From corporate city to citizens city?: Urban leadership *after* local entrepreneurialism in the U.K. *Urban Affairs Review, 35,* 3–23.

Henderson, J. (2004). The politics of mobility and business elites in Atlanta, Georgia. *Urban Geography, 25,* 193–216.

Heynen, N. (2006). Green political ecologies: Towards a better understanding of inner-city environmental change. *Environment and Planning A, 38,* 499–516.

Hughes, R. (1992). *Barcelona.* London: Harvill Press.

Jessop, B. (1997a). A neo-Gramscian approach to the regulation of urban regimes: Accumulation strategies, hegemonic projects, and governance. In M. Lauria (Ed.), *Reconstructing urban regime theory: Regulating urban politics in a global economy* (pp. 51–73). Thousand Oaks, CA: Sage.

Jessop, B. (1997b). The entrepreneurial city: Re-imaging localities, redesigning economic governance, or restructuring capital? In N. Jewson & S. MacGregor (Eds.), *Transforming cities: Contested governance and new spatial divisions* (pp. 28–41). London: Routledge.

Jessop, B., Peck, J. A., & Tickell, A. (1999). Retooling the machine: Economic crisis, state restructuring, and urban politics. In A. E. G. Jonas & D. Wilson (Eds.), *The urban growth machine: Critical perspectives two decades later* (pp. 141–159). Albany: State University Press of New York.

Jonas, A. E. G., & Ward, K. (2002). A world of regionalisms? Towards a US–UK urban and regional policy framework comparison. *Journal of Urban Affairs, 24*(4), 377–401.

Jonas, A. E. G., While, A. M., & Gibbs, D. C. (2004). State modernization and local strategic selectivity after Local Agenda 21: Evidence from three northern English localities. *Policy and Politics, 32*(2), 151–168.

Jonas, A. E. G., & Wilson, D. (Eds.). (1999). *The urban growth machine: Critical perspectives two decades later.* Albany: State University Press of New York.

Judd, D., & Ready, R. (1986). Entrepreneurial cities and the new policies of economic development. In G. Peterson & C. Lewis (Eds.), *Reagan and the cities* (pp. 209–242). Washington, DC: The Urban Institute.

Keil, R., & Graham, J. (1998). Reasserting nature: Constructing Urban environments after Fordism. In B. Braun & N. Castree (Eds.), *Remaking reality: Nature at the turn of the millennium.* London: Routledge.

Keil, R. (2005). Progress report—urban political ecology. *Urban Geography, 26,* 640–651.

Krueger, R., & Savage, L. (2007). City-regions and social reproduction: A "place" for sustainable development? *International Journal of Urban and Regional Research, 31,* 215–223.

Lake, R. W. (2000). Contradictions at the local scale: Local implementation of Agenda 21 in the USA. In N. Low, B. Gleeson, I. Elander, & R. Lidskog (Eds.), *Consuming cities: The urban environment in the global economy after the Rio Declaration* (pp. 70–90). London: Routledge.

Lake, R. W. (2003). Dilemmas of environmental planning in post-urban New Jersey. *Social Science Quarterly, 84,* 1003–1017.

Lauria, M. (Ed.). (1997). *Reconstructing urban regime theory: Regulating urban politics in a global economy.* Thousand Oaks, CA: Sage.

Logan, J., & Molotch, H. L. (1987). *Urban fortunes: The political economy of place.* Berkeley: University of California Press.

Marshall, T. (1992). Industry and environment in the Barcelona region. *Town Planning Review, 63,* 349–364.

Marshall, T. (1993). Environmental planning for the Barcelona region. *Land Use Policy, 10,* 227–240.

Marshall, T. (Ed.). (2004). *Transforming Barcelona.* London: Routledge.

Mayer, M. (1994). Post-fordist city politics. In A. Amin (Ed.), *Post-Fordism: A Reader.* Oxford, UK: Blackwell.

McCann, E. J. (2002). The urban as an object of study in global cities literatures: Representational practices and conceptions of place and scale. In A. Herod & M. Wright (Eds.), *Geographies of power: Placing scale.* Oxford, UK: Blackwell.

McManus, P. (2004). *Vortex cities to sustainable cities: Australia's urban challenge.* Sydney: University of New South Wales Press.

McNeill, D. (1999). *Urban change and the European left: Tales from the new Barcelona.* London: Routledge.

Meadowcroft, J. (2005). From welfare state to ecostate? In J. Barry & R. Eckersley (Eds.), *The state and the global ecological crisis* (pp. 3–24). Cambridge, MA: MIT Press.

Mendes, W. (2005, Fall). *Creating a "just and sustainable" food system in the City of*

Vancouver: The role of governance, partnership and policy-making. Unpublished PhD dissertation, Department of Geography, Simon Fraser University, Burnaby, BC.

Molotch, H. L. (1976). The city as a growth machine: Toward a political economy of place. *American Journal of Sociology, 82,* 309–330.

Montalbán, M. V. (2004). *Barcelones.* Barcelona: Editorial Empuries. (Originally published in 1987)

Mumford, L. (1961). *The city in history.* London: Penguin.

Neiman, M., & Loveridge, R. (1981). Environmentalism and local growth control: A probe into the class bias thesis. *Environment and Behavior, 13,* 759–772.

Nello, O. (2004). Urban dynamics, public policies and governance in the metropolitan region of Barcelona. In T. Marshall (Ed.), *Transforming Barcelona* (pp. 27–46). London: Routledge,.

O'Connor, J. (1998). *Natural causes: Essays in ecological Marxism.* New York: Guilford Press.

Organization for Economic Cooperation and Development (OECD). (2001). *Cities for citizens: Improving metropolitan governance.* Paris: Author.

Peck, J., & Tickell, A. (2002). Neoliberalizing space. *Antipode, 34,* 380–404.

Pincetl, S. (2003). Nonprofits and park provision in Los Angeles: An exploration of the rise of governance approaches to the provision of local services. *Social Science Quarterly, 84,* 979–1002.

Platt, R. (2006). Urban watershed management: Sustainability, one stream at a time. *Environment, 48,* 26–42.

Qushair, G. (2006). Joan Clos: Former mayor of Barcelona. Accessed October 2006 at *www.citymayors.com/mayors/barcelona_mayor.html.*

Raco, M. (2005). Sustainable development, rolled-out neoliberalism and sustainable communities. *Antipode, 37,* 324–347.

Rees, W. E. (1992). Ecological footprints and appropriated carrying capacity. *Environment and Urbanization, 4,* 121–130.

Satterthwaite, D. (1997). Sustainable cities or cities that contribute to sustainable development? *Urban Studies, 34,* 1667–1691.

Scott, A. J., & Storper, M. (2003). Regions, globalization, development. *Regional Studies, 37,* 579–593.

Stone, C. N. (1989). *Regime politics: Governing Atlanta, 1946–1988.* Lawrence: University of Kansas Press.

Stone, C. N. (1993). Urban regimes and the capacity to govern: A political economy approach. *Journal of Urban Affairs, 15,* 1–28.

Swyngedouw, E. (1996). The city as a hybrid: On nature, society and cyborg urbanization. *Capitalism, Nature, Socialism, 7,* 65–80.

Swyngedouw, E., & Heynen, N. C. (2003). Urban political ecology, justice and the politics of scale. *Antipode, 35,* 888–918.

Tello, E. (2004). Changing course?: Principle and tools for local sustainability. In T. Marshall (Ed.), *Transforming Barcelona* (pp. 225–250). London: Routledge.

While, A., Jonas, A. E. G., & Gibbs, D. C. (2004a). The environment and the entrepreneurial city: Searching for a "sustainability fix" in Leeds and Manchester. *International Journal of Urban and Regional Research, 28,* 549–569.

While, A., Jonas, A. E. G., & Gibbs, D. C. (2004b). Unblocking the city: Growth pressures, collective provision and the search for new spaces of governance in Greater Cambridge, England. *Environment and Planning A, 36,* 279–304.

Worster, D. (1985). *Rivers of empire: Water, aridity, and the growth of the American West*. New York: Pantheon.

Worster, D. (1994). *Nature's economy: A history of ecological ideas*. Cambridge, UK: Cambridge University Press.

Zimmerer, K. S. (2000). The reworking of conservation geographies: Nonequilibrium landscapes and nature–society hybrids. *Annals of the Association of American Geographers, 90*, 356–369.

CHAPTER 6

Integrating Sustainabilities in a Context of Economic, Social, and Urban Change

The Case of Public Spaces in the Metropolitan Region of Barcelona

MARC PARÉS
DAVID SAURÍ

The concept of sustainability too often has been approached either from a strict environmental logic or from a neoliberal view in which capital accumulation is seen as the way to solve socioenvironmental problems. However, and despite the Rio agreements of 1992, the social dimension of sustainability has received much less consideration, while the political dimension, grounded in the involvement of the citizenry in the construction of a sustainable future for their communities, has been virtually ignored.

In this chapter we propose a reinterpretation of the sustainability concept, emphasizing the social and political dimensions. We also attempt to integrate the different dimensions of sustainability at two levels: transversal and scalar. Our analysis is situated in the context of the important social, economic, and territorial transformations currently occurring in advanced societies, which we examine for the Metropolitan

Region of Barcelona (MRB). The chapter focuses on public spaces in the MRB in order to draw out some theoretical insights regarding the sustainability of different urban forms (compact or high-density, and dispersed or low-density), the number and characteristics of public spaces present in these urban forms, and the life styles they tend to foster.

From the analysis of public spaces, we want to test whether the new social and urban model emerging in the Barcelona region is a sustainable one, not only from an environmental standpoint but also from the economic, and especially the social and political, dimensions of sustainability (Valentin & Spangenberg, 2000). Our aim is therefore to produce an integrated assessment of urban sustainability (Ravetz, 2000; Rotmans, Asselt, & Vellinga, 2000) stressing the role of public spaces in cities, asking whether these public spaces incorporate sustainability components, and identifying what these spaces can contribute to advance urban societies toward sustainability goals.

More precisely, the main objective of this chapter is to analyze whether public spaces are spaces of sustainability, using two central rationales. The first, of a distinct social and political nature, is that public spaces may generate social capital and thus facilitate new forms of participatory governance. In this sense public spaces may improve sustainability by increasing social cohesion and civic engagement. The second argument, along environmental lines, interrogates public spaces on the use of local species and efficient irrigation practices for their gardened areas.

The analysis of public spaces of the MRB is framed in the current urban and social context that is characterized by a change of the urban form (expansion of sprawl) and also by social changes derived from new norms and behaviors characteristic of a high social heterogeneity and a strong process of individualization. As in other parts of the Western world, a new type of society is emerging in the MRB that requires new forms of government that are closer to the people.

The new—postmodern for some—societal model and its associated geographical development has conditioned the creation of public spaces in the cities. These spaces become more common in the compact, high-density areas, whereas land privatization dominates in the low-density areas of urban sprawl.

Following this outline of the central arguments of the chapter, there are two important paradoxes that must be confronted from an integrated perspective on sustainability. First, in a time when repeated calls for public participation and new forms of governance are appearing, in the MRB we can see a form of urban development (i.e., urban sprawl) largely deprived of public spaces and consequently less prone to stimulate the role of such spaces in promoting social capital and public interaction. The second paradox, this time from an environmental perspective, is that public spaces developed in the sprawl have smaller impacts (for example, in terms of water consumption) than public spaces created in the high-density areas. It seems therefore, that environmental, social, and political sustainability drift apart in the growing spatial polarization of the urban form in the Barcelona region.

The chapter is structured as follows. After this introduction, which includes also a brief note on our study area and on the methodology used in the research, we develop our theoretical framework on integrated sustainability, which is then applied to public spaces. The third section deals with the social and territorial transformations brought about by the current historical period in our study area. This is followed by an attempt to integrate sustainability at two levels: scalar (in reference to the two urban forms, dispersed and compact), and transversal, the latter presented in section 5. Finally, we draw some conclusions in which we show how public spaces in the compact areas of the Barcelona region may foster social and political sustainability despite their relatively poor environmental performance.

STUDY AREA, MATERIALS, AND METHODS

The Metropolitan Region of Barcelona includes seven *comarques* (counties) and 163 municipalities (see Figures 6.1 and 6.2). Internally, it can be divided into one core area (the city of Barcelona), a first periphery, a second periphery, and six cities with more than 50,000 inhabitants (1996) that act as subcenters. The first periphery coincides with the urban continuum of the city of Barcelona and, as the capital of Catalunya, it is characterized by high urban densities. The second periphery, on the other hand, is based on low-density settlement patterns. Since approximately the beginning of the 1980s, people, residences, and jobs have moved increasingly from the metropolitan compact core (Barcelona and the first periphery) to the low-density outer areas (second periphery) (Nel·lo, 2001).

According to Garbancho (1995), public spaces in Barcelona added up to about 30 hectares at the beginning of 20th century. They increased to 100 hectares during the Second Spanish Republic (1930s) and to 130

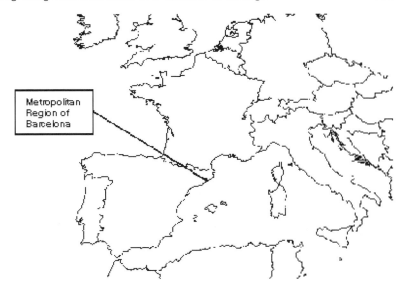

FIGURE 6.1. Location map of the Metropolitan Region of Barcelona.

hectares during the Franco regime, a period in which the population of the MRB grew by 1 million people. From the late 1970s until the Olympic Games of 1992, urban public spaces expanded to cover around 800 hectares. With the important urban developments of the 1980s and early 1990s linked to the Olympic Games, the city of Barcelona gained new public spaces, making the city less dense and more habitable than in the past. However, other questions affecting public spaces in the metropolitan area appear important as well, especially whether public spaces have increased in the suburban locations along with population and economic activity.

This research has used what we call a "methodological triangulation." By this expression we refer to the integration of quantitative and qualitative methodologies, leading toward an integrated assessment of the sustainability of urban public spaces.

One component of the research, the quantitative work, involved producing the digital cartography of urban gardened public spaces of the MRB, developing a numerical analysis of several variables from a sample of 315 of these public spaces, and sending a questionnaire to the 163 municipalities of the MRB on the various characteristics of these spaces. Since water is a critical resource in this area, we have chosen this parameter as a means to examine environmental sustainability practices. The

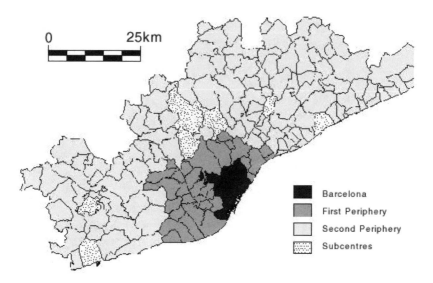

FIGURE 6.2. The Metropolitan Region of Barcelona.

second component of the research involved, qualitative work, was based on 38 open but detailed interviews with residents of the region and on 19 direct observations of urban public spaces in this area.

The 38 persons interviewed were selected based on the following criteria:

- The predominant urban form of their municipality (i.e., whether compact or dispersed).
- Social capital (i.e., whether a member of a voluntary organization or not).
- Social class (based on the average income per capita of their municipality).
- Sex.
- Age.

Bearing in mind the results of the quantitative methodology and the differences identified between compact and dispersed cities in what concerns the presence of public spaces, the interviewers were directed to explore the following issues:

- Urban lifestyles associated with different urban models.

- The role of public space in the generation of social capital and social cohesion.
- The role of public space in the generation of participatory political cultures.

The third methodological "leg" involved 19 direct observations of urban public spaces in several municipalities within the MRB. The main criteria for selecting these public spaces was the urban form surrounding the public area (i.e., whether compact or dispersed) and the characteristics of the specific municipality. Each selected space was observed during a 60-minute period at similar hours and days of the week.

INTEGRATED SUSTAINABILITY: INCORPORATING THE SOCIAL AND POLITICAL DIMENSIONS

Despite the wide recognition of the multidimensionality of the sustainability concept, in practice there has been a significant emphasis on the environmental dimension and a relative neglect of the other dimensions, most notably the social (Adams, 1990). The environmental dimension of sustainability refers to the need to preserve the natural assets of the earth so that they are available for future generations. In other words, the objective of sustainability from the environmental perspective consists in guaranteeing that the current model of societal development will proceed without jeopardizing critical natural capital for the next generation. The ensuing debate has offered a variety of interpretations regarding the significance of critical natural capital. Neoliberal discourses based on the logic of capitalist development and the discourse currently guiding many governmental actions deny any incompatibility between economic growth and the preservation of natural capital under the well-known axiom that exhausted resources may always be compensated by technological innovations (Gowdy & O'Hara, 1997; Mas-Collell, 1994). For their part, political-ecological views express a blatant criticism of these arguments, since they assert that there exists a fundamental contradiction between capitalism and environmental conservation (Robbins, 2004).

However, the social dimension of sustainability remains much less examined, both at the theoretical and applied levels. Furthermore, when this dimension is the object of analysis, integration with its environmental and economic counterparts is usually altogether absent. Social or sociocultural sustainability may be understood as one that pursues the stability of social and cultural systems or that guarantees the durability of social capital within a particular society. The objective therefore is to maintain the interactions between individuals and social networks as well as the trust and reciprocity norms that arise from these interactions (Putnam, Leonardi, & Nanetti, 1994). Using a welfare approach, Garcés, Ródenas, Sanjosé, et al. (2003) define social sustainability as an extension of the principle of social welfare through time, understanding that welfare is a right not only of current generations but also of the citizens of the future. Therefore, the sustainable path of social systems requires equity and social justice, both for the present and for the future (Burton, 2001). That said, integrated sustainability from the social standpoint may be interpreted as the goal of ensuring the durability over time of the citizen's quality of life. This requires the preservation of sound economic and environmental conditions, but, above all, it demands the welfare of citizens. Only by incorporating the social dimension into the sustainability concept can we expect the concept to fulfill its potential as a tool to allow for community development and social welfare in harmony with the environment and based on cohesion and social justice.

Departing from this approach, Hediger (2000) proposes incorporating the economic, environmental, and social components of sustainability into an integrated model and resituates the debate between "weak" and "strong" versions of sustainability, in which the social dimension as well as the environmental dimension are introduced as fundamental requirements for the strong sustainability view. This may constitute a good theoretical point of departure for the definitive integration of the social component within sustainability. On the other hand, it is also true that the success of sustainable development depends also on its ability to attract society to this model of development (Barr, 2003; Burgess, Harrison, & Filius, 1998). Environmental impacts cannot be reduced unless the citizenry is fully aware of their significance (Macnaghten & Jacobs, 1997), and the citizenry will not become aware without being implicated in the process and without considering the economic and social elements of the new development model. Hence, a sustainable model cannot be thought of solely in environmental terms but rather in more comprehensive ways; otherwise, it may be condemned to failure.

Some authors (see, e.g., Spangenberg & Bonniot, 1998; Valentin & Spangenberg, 2000) have introduced the concept of institutional sustainability to refer to the organizational and procedural mechanisms needed to advance the compromise of a society toward sustainable development. We reformulate this view in terms of *political sustainability*, arguing that there is not a universal form of sustainability for all places but rather that each place has to define its own development model based on the will of its citizens.

The political dimension of sustainability is somewhat different from the three dimensions already commented upon in that it is not substantive in character but rather it is a dimension that deals with how to achieve, from public policies and public decision making, the goals of sustainable development. Nevertheless, political sustainability is extremely important for at least two reasons. First, because in our current society, complexity, uncertainty, the diversity of social actors, and the interrelationships between the different sustainability dimensions make it necessary that decision making be dispersed as widely as possible, incorporating all the voices at stake. In this sense, Adger et al. (2003) explain that environmental decisions necessarily require political legitimation. In other words, integrated sustainability must be grounded in a form of governance that institutionally legitimizes the decisions made. Second, most elements of a sustainable society require a strong involvement by the citizenry that must be complicit with (i.e., invested in) this new model of development and act accordingly in their own attitudes and behavior (Macnaghten & Jacobs, 1997). Since sustainability depends on societal ability to guide the nature–society interface through more sustainable trajectories (Kates et al., 2001), the social compromise becomes of key importance.

In sum, we can conclude that the version of sustainability leading toward social change will be one developing from the logic of integration of the environmental, economic, social, and political dimensions. This integration should lead toward a real social transformation with the active participation of the citizenry. This is why it is so important to introduce the social and political aspects within sustainability discourses (see Figure 6.3).

Integrated Sustainability of Public Spaces

We now turn our attention to urban public spaces in the Metropolitan Region of Barcelona, using the insights of integrated sustainability assessment. First of all, this integration must be an integration of scale. It is necessary to understand public spaces in their economic, social, and territorial contexts. Therefore, an analysis of their sustainability condition cannot be approached from a strict vision of each singular public area,

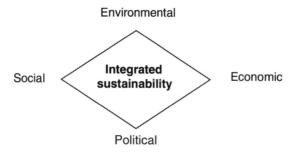

FIGURE 6.3. Integrated sustainability.

but rather we must take into account where this space is located (scale) and also the wider framework of societal change occurring in the region (transversality).

In this respect, we must mention that a series of social, economic, and territorial transformations has triggered changes in the number and characteristics of public spaces in the MRB. First, global and local spheres coincide in this area. Both also coincide, and even under a more pronounced form, in what concerns public spaces. Hence, we must understand these spaces in the framework of a new social and territorial scenario that permeates down to the local scale.

Second, sustainability integration must be approached transversally, taking into consideration the social, environmental, economic, and political arenas that intervene in the process of the creation of public spaces. It is not possible to analyze these areas solely from an economic or environmental point of view. If a society must become sustainable, it must become so in its entirety. In this sense, then, particular sustainable pathways from an environmental perspective must not collide with social, economic, or political pathways. It is from this integrative approach that we must ascertain whether public spaces are sustainability elements, what their contribution is toward a more sustainable society, and what the requirements for a public space to become sustainable really are.

From an environmental perspective, in a Mediterranean area such as the MRB, one possible way to assess the sustainability of public spaces is to examine water use and management in the design (species planted) and maintenance (irrigation practices) of the gardened areas. Hence, a public space in the Barcelona region will be more environmentally sustainable insofar as it minimizes water consumption and contains species well adapted to the Mediterranean climate.

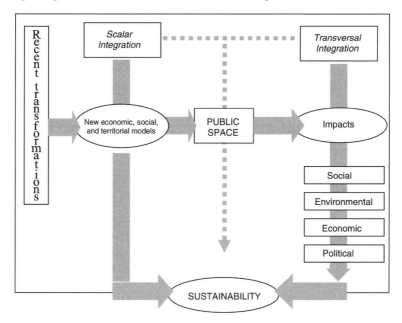

FIGURE 6.4. Integrated assessment of public spaces.

From the social side, public spaces may become sustainable assets as long as they contribute to a greater social cohesion of the urban environment in which they are located, hence increasing the level of social capital of the community that uses these areas. Therefore, a public space may be sustainable if it serves community social relations and acts as an element that expresses the heterogeneity and diversity of a community. On the other hand, however, a public space may turn into a socially unsustainable space if it turns out to be a space of exclusion, fear, or social segregation (Fenster, 2004; Marne, 2001; Massey, 1994; Pain, 2000).

Focusing now on the economic dimension, a public space may serve as a tool for economic development if it articulates the place in which it is located, for example, attracting business and other economic activities. But it may be converted into a source of segregation if it is privatized and access is restricted to minimal collective uses. Equally, but conversely, an excessive focus on consumption may diminish its role as a space that encourages social relations.

In political terms, public spaces may act as catalysts of participation and the involvement of the citizenry in all that is public. Likewise, public spaces may appear as building blocks of a more democratic society to the extent that they nurture public expression and public demands. In this sense, public spaces are not only places for political expression buy also spaces for social and cultural discourses in which the community's diversity may manifest itself in an open way.

TRANSFORMATIONS OF "SECOND MODERNITY" AND THEIR IMPACT ON PUBLIC SPACES

During recent decades a number of economic, social, and territorial transformations have left a large imprint on Western societies. In this section we will summarize the main characteristics of the model emerging from these changes in general, and more specifically for the MRB. Afterward, we will examine how these transformations may affect public spaces and their sustainability in our study area.

A New Socioeconomic Model

Current western societies are ruled by a new form of capitalism that offers some peculiarities with respect to previous forms based more on mass production and mass consumption. The change from Fordism to post-Fordism is a key element in understanding this new face of capitalism. Thus, a system that generated goods and services for homogeneous social groups has been replaced by a new system in which more specific goods and services are put into the market, thus offering larger choices for the individual consumer, allowing her or him "to be different" vis-à-vis other consumers. Mass consumption becomes more diversified and consumers may have more preferences, although all of them intentional (Bauman, 2001).

The most significant change brought about by the post-Fordist system, with devastating effects on labor organizations, has been the fragmentation of the labor force, eroding collective links and solidarities among workers. The deregulation of production and financial services is one of the most significant features of globalization, but it would be a mistake to think of globalization in just economic terms. Globalization also exerts very important effects in places and individuals, and in the relationships established between them (Giddens, 2000). Moreover, their consequences are very different, depending on the specific social group. Place occupies a central role in the circulation of people and capital in a globalized world, since globalization occurs through specific economies and societies also linked to specific places (Sassen, 1998).

The different spatial scales (local, regional, global) have become inextricably interwoven under a hierarchy clearly determined by globalizing processes that, in turn, produce crucial geographical differences (Harvey, 2003). Under this context, social, economic, and cultural issues become more and more complex, and their analysis requires a clear understanding of the relations of the local with the global (Swyngedouw, 1997).

Transformations recently carried out in Western societies are not only economic but also social and political. Beck, Giddens, and Bauman, among others, argue that the modern industrial society has transformed itself into a "second modernity" characterized by trends such as individualization, globalization, and the generation of risks and uncertainties (Bauman, 1992, 2001; Beck, 1998, 2002; Giddens, 1999, 2000). This second modernity has changed institutional organizations such as social classes, the family, patriarchy, labor stability, technical certainty, the control of nature, etc. All the above must confront a much more complex society in which the individual gains prominence. In this new social model, social structures are more heterogeneous and are anchored on new inequality forms. Families have become more diverse, the distribution of roles has also changed, and, at the same time, new "postmaterial" values have appeared.

This is therefore a society with new risks and uncertainties and one more difficult to govern with traditional institutions. At the same time, the nation-state loses momentum to the benefit of the global–local nexus. In the local agendas, new issues (environmentalism, feminism, pacifism, etc.) are gaining terrain, and new actors and social movements expand incessantly. These actors and movements may think locally, but the repercussions of their actions are global.

New issues, actors, and social movements force us to think in new forms of government and in new public policies at the local level. In the face of the complexity and uncertainty of our current society, governance, relational and network government based on proximity, and new mechanisms for public participation are needed to incorporate social agents into decisions that are complex, less certain in scientific and technical terms, and, above all, more prone to conflict than before (Blanco & Gomà, 2002; Brugué & Gomà, 1998; Stocker, 1996; Subirats, 1997). In parallel, however, the same characteristics that current societies tend to harbor are new, more private urban lifestyles in which, as we will see later, citizens look inward to their residences and families, search for privacy, and choose to become anonymous to the rest of the community.

The New Urban Model of the MRB: Banal, Dispersed, Segregated

Cities are perhaps the best example of the large economic and social transformations at the global scale that have been occurring in recent decades. The postmodern city is becoming transformed by several processes, all of them aligned in the same direction of homogenization, inequality, and lacking a collective identity. The internationalization of economic fluxes and the globalization of information technologies have increased the importance of large urban centers. At the same time, however, we are seeing a process of economic and territorial deconcentration whereby those economic sectors with higher added value remain in the city cores while the less valuable are expelled toward the metropolitan peripheries.

Under this pattern, the loss in productive activities has been compensated for by a tertiarization of metropolitan centers, which have increasingly specialized in the service economy and have strongly oriented themselves toward leisure and consumption. New public spaces are generated, but with a strong private character, an emphasis on security, and on restricting uses and access (Bridge & Watson, 2000; Zukin, 1998).

This process is especially evident these days in Mediterranean cities such as Barcelona, where the traditionally compact urban form is quickly losing its character to the benefit of more dispersed urban land uses reflecting the characteristics of the more Anglo-Saxon phenomenon of urban sprawl (Albet & Riera, 1998; Nel·lo, 2001). Urban sprawl occurs simultaneously with the revitalization of urban centers by higher-income groups that have "rediscovered" the appeal of living in city cores and are gentrifying traditional neighborhoods. In Barcelona, for instance, gentrification is producing a new urban space that is often elitist and oriented toward consumption, tight security, and global tourism and that ends up creating trivial or banal settings (Muñoz, 2004).

Centrality and dispersion are therefore two complementary and not opposed trends occurring in major metropolitan areas, including that of the Metropolitan Region of Barcelona. In the latter and during the past 20 years, this territorial model has suffered important changes, most notably, and as noted earlier, through the phenomenon of urban sprawl.

Thus, segregated residential, productive, and, to a lesser extent, leisure activities have relocated toward the outer part of the first periphery and especially toward the second periphery, where low-density form is gaining presence with a strong growth in single and semidetached housing. This is accompanied by internal migratory shifts by which citizens are moving out of the central core and are occupying the municipalities of, especially, the second periphery. Overall, the total population of the MRB has remained fairly stable during the past two decades, but new urban land has increased significantly (more than 20 percent between 1993 and 2000, for example), mostly to accommodate internal residential migrants (the city of Barcelona lost almost 400,000 people between 1981 and 2001).

Also increasingly, the common functions of the city are not being satisfied by the citizenry in the municipalities where they live. The emerging dispersed low-density model proliferates everywhere without any coherence other than proximity to the main communication networks (mostly roads and highways). In our view, this is a model that does not generate "citizenship," since identity (i.e., the sense of belonging to a community) and the most significant community relationships become totally blurred.

The transformations described in the previous sections also have deep implications for sustainability, both at a global scale and also at the scale of public spaces. From sustainability logic our objective is to link the new global "second modernity" of economy and society with the specific public spaces located in an area with different urban forms and with citizens whose lifestyles are also different.

In the following sections we will analyze the impacts of the transformations induced by this second modernity on the MRB in general and how they affect specifically the public spaces of this area.

SCALAR INTEGRATION

To analyze the impacts of recent social and territorial transformations in the MRB, we will use two kinds of integration, the scalar and the transversal. Dealing first with scalar integration, we focus on two issues: first, how these transformations have generated new urban lifestyles in the MRB, with important implications for sustainability; and second, how these economic, social, and territorial transformations and associated changes in urban lifestyles are influencing public spaces.

New Urban Lifestyles

First of all, the new urban lifestyles, in themselves very different from those prevalent in the more industrial society of the past, are not unique or homogeneous. Rather, they are multiple and diverse—and here we will defend the idea that these new lifestyles are intimately connected with the different urban forms present in the MRB. Also, each of these emerging lifestyles bears different impacts in terms of environmental, social, political, and economic sustainability.

The proximity to most services, the easiness in satisfying most needs, and the variety of services provided by the compact city are the main reasons argued by citizens living in these urban areas to justify their choice of residence. The compact form then is characterized by diversity and activity: there is interaction, exchange, and relationships. Therefore, density tends to cultivate complexity and diversity. The following sentences, extracted from the interviews, illustrate and reinforce our assertions:

"People that live in the neighborhoods . . . can live there, and it is not necessary to go elsewhere, because these neighborhoods have good services—not only schools and hospitals but also a lot of shops, bars, coffees. . . . My job is near my home, just 5 minutes away. I can do everything here, it is fantastic!" (adult woman, member of a voluntary organization, from a low-income municipality and compact urban form)

"If you live in company, in a community, your life is more secure. Here you know neighbors of this building and all of the street. It is very nice to have relationships with other people, to talk." (old man from a low-income municipality and compact urban form)

"A good quality of life, for me, is to have good services near here—that's what we have in the city." (adult woman from a low-income municipality and compact urban form)

"I feel very well here because of diversity. You can go to the theater, the cinema. . . . The city has a very big supply of activities." (young man, member of a voluntary organization, from a low-income municipality and compact urban form)

The values and behavior of residents in the low-density or dispersed city are much more different. These citizens are fully aware of the limitations that are entailed by living in the low-density suburbs, but they also acknowledge two elements that compensate for these limitations: tranquility and privacy. In this sense, the urban low-density model promotes an individualization process based on an increase of privacy and lacking a stimulus for the diversification of collectivities. The following statements extracted from the interviews illustrate this fact.

"Here I have tranquility, I have a lot of open space around my home. Life is very different here. There is no stress." (old woman from a low-income municipality and dispersed urban form)

"This is no place to find people. To walk we go to the city center and not around here; houses are single-family units, there is no community, there is no neighborhood life, everything is quiet around here." (adult woman, member of a voluntary organization, from a low-income municipality and dispersed urban form)

"In my street you cannot see anybody. Nobody walks here. It is a residential zone. . . . There is no relation between people. My relationships are with my own family and that's all." (adult woman, member of a voluntary organization, from a high-income municipality and dispersed urban form)

Our argument is supported from different angles. First, the forced mobility of many citizens of the low-density areas implies the use (and abuse) of the private car. This considerably limits social interactions in the urban space. Second, because many of the functions characteristic of a city cannot be accomplished in the low-density areas, mobility becomes the rule for most uses. Third, the strict residential character of these urban areas implies that they lack services, shops, leisure activities, and other spatial spheres of relation. These are precisely the elements that spawn complexity and that are inherently urban in nature.

The forced mobility of the low-density model not only stimulates individualism and limits relationships among people, but it also contributes enormously to energy and pollution emissions due to the overwhelming dependency on private vehicles (Tello, 2000). These are not the only significant environmental impacts of the low-density model (see also Barr, 2003; Speir & Stephenson, 2002). Regarding water, for example, we have observed that in the MRB per-capita consumption in the denser metropolitan core totals some 130 liters per day whereas in the low-density second periphery average consumption jumps to 275 liters per capita per day (Saurí, 2003; Saurí, Capellades, Rivera, & Paredes, 2001). Finally, the low-density model also bears such important effects as ecological fragmentation and increases in some environmental hazards, such as floods or forest fires (Badia, 2000; Rueda, 1995; Tjallingii, 2000).

Single houses, many of them isolated and endowed with a private garden, are thought to cherish tranquility and to promote living "inward." Thus a lifestyle addressed to escaping from collectivity and from social relations in order to live privately becomes satisfied. All together, this produces enormous obstacles to the creation of a sense of identity with the neighborhood of residence, since the links generated are minimal and the lack of social ties makes the construction of a sense of belonging to a place very difficult. Low-density urbanism therefore minimizes relations, attenuates spontaneous contacts and the generation of links, and plagues the creation of collective sensitivities with problems.

Nevertheless, privacy and social anomie exist also in the compact city, where the presence of many more people combines with the effects of individualization. In the populous core of the MRB it is difficult for citizens to know one another and to engage in a civic compromise with the city. Even so, the role of neighborhoods remains crucial to achieve a sense of identity and a greater civic involvement. Hence, in the compact city we find many neighborhoods that have developed an important community identity, embracing the citizenry and creating valuable social capital through strong associative networks and interpersonal trust.

The Recent Transformation of Urban Public Spaces

In order to analyze the territorial, economic, and social transformations affecting public spaces in the MRB, we must begin by observing their spatial distribution. Thus, we have selected two indicators that measure respectively the relative area of gardened public spaces (percentage of gardened public space with respect to total urban land area) and gardened public space per person (see Table 6.1).

Both indicators show how the gardened public space diminishes from the compact to the low-density city. There are many variables that contribute to explain this variation, but urban density achieves the high-

TABLE 6.1. Gardened Public Space in the MRB

	Area in 1997 (hectares)	Percentage of total gardened area	Percentage of total urban land	Square meters of garden per person
Barcelona	434.6	43.4	5.81	2.88
First periphery	310.9	31.0	2.42	2.22
Subcenters	136.2	13.6	2.19	2.15
Second periphery	119.8	12.0	0.50	1.75

est explanatory power. Thus, a decrease in density produces a parallel decrease in gardened public space almost linearly. Low-density urbanism is characterized by single houses with private gardens and at the same time by a smaller quantity of public gardened spaces. Therefore, in these areas the private green space advances clearly over the public green space. The larger presence of single housing in the low-density areas facilitates the creation of personal spaces for individuals, and this is an asset much in demand by current society. This personal space is not only indoors but also outdoors, since many houses have gardens as well. Thus, the presumed higher quality of life provided by greenness finds its way into the individual realm. Moreover, private gardens, especially those exhibiting lavish grass surfaces, are considered as "positional goods" (Hirsch, 1980), that is, a good whose value comes from owning something that others do not have.

In general, then, the low-density urban model shows an increase in the size of private spaces and a decrease of public spaces. Not only is the public gardened area smaller, but the public and collective spaces in general are also smaller. The traditional public square, a classic meeting point for the citizenry, is also very scarce. This trend reflects the fact that there is a smaller need to endow the city with public spaces, since the individual gets preference over the group, and therefore the collective space loses significance as compared to the individual space. Collective demands are geared toward individual autonomy, and the demand for public spaces ceases to be a priority.

On the other hand, the lesser need to offer public spaces is explained, because the low-density model provides at the individual level and through private gardens what the compact city offers at the collective level—that is, public gardens. In the context of individualization, we can conclude that there exists a certain coherence between the model of social relations and the urban model. In this sense, we can speak of an individualization of the low-density city that spawns privacy and is deprived of public spaces and an individualization of the compact city, where a model with more public space gives room to new social relations, individualized but diversified at the same time.

This distinction between individualization in the compact city and individualization in the dispersed city does not imply that individuals in either model are necessarily subject to different individualization processes, despite the fact that the social relations implicit in each urban model may influence these processes. The despatialized character of our current society, where relations tend to drift away from particular geographical locations, leads to the fact that relationships between the two models intertwine through individuals. This gives space to individualization processes connected to each individual.

All that said, it is also true that the compact city may not develop an urban setting that fosters diversity and social integration. Thus, we can find compact urban models that fragment and segregate through urban designs zoned not just functionally but also along social, ethnic, or cultural lines. This is precisely the distinction that Beck elaborates between the city of "and" and the city of "or." The latter is the city of segregation, of separation, and of social disintegration. At the opposite site, the city of "and" is that designed to stimulate diversity produced by individualization and to integrate this diversity into the collective space, thus generating strong social identities.

The city of "and" is possible in the compact urban form when, through the means provided by public spaces, on occasion these social identities may be generated. It is therefore in this urban form with more collective space where individualization may develop its social component. On the contrary, in the dispersed city model, the creation of social identities remains blocked by the lack of such spaces. Social identities, then, do not arise in private spaces but in collective spaces, and this is precisely what is lacking in the low-density model.

TRANSVERSAL INTEGRATION: THE ENVIRONMENTAL, SOCIAL, AND POLITICAL IMPACTS OF PUBLIC SPACES

Following our review of how social, economic, and territorial transformations have generated new urban lifestyles and have affected the role and distribution of urban public spaces in the MRB, we now turn our attention to the integration of sustainability in public spaces by analyzing their environmental, social, and political impacts in the context of global trends.

Environmental Sustainability: The Environmental Paradox of Public Spaces in the MRB

In order to examine the environmental performance of urban public spaces in the MRB, we will select a single vector—water—and sharpen our analysis on environmental sustainability through the critical assessment of water use, species planted, and Xeriscaping (sophisticated water-conservation) practices (Burés, 2000).

After calculating the water consumption of public gardening in the MRB, we can conclude that it is not a very relevant factor in the total water use of the area. Thus, the approximately 7.6 cubic hectometers consumed by this type of gardening represent only about 1.5% of the total water consumption in the Barcelona region, and about 4% of total domestic uses (Parés, Domene, & Saurí, 2004). However, the internal migratory shifts from the metropolitan core to the second periphery, especially in the predominant form of single and semidetached housing, is feeding the growth of private gardens. In 2001, approximately 8% of total water consumption in the MRB, and about 18% of domestic uses, could be attributed to private garden irrigation (Domene & Saurí, 2003). To the extent that the low-density city is expanding, this figure could increase in the future, thus adding new stresses to the already precarious equilibrium between resources available and water demand in the study area.

The impact of public and private gardening upon water consumption in the MRB not only depends on the total land area devoted to gardening but also on the type of garden designed and the management practices undertaken. In this sense, the public sphere often acts as a possible model for imitation by the private sphere, since many individuals inform their attitudes through imitation either from fellow citizens or from public activities. Hence, public gardening may play an important role in developing a model to be followed by private gardens, both in the design and type and in management practices.[1]

As to garden types, in the MRB we found a clear dominance of public gardens with high water consumption. The smaller municipalities, where the low-density model is more widespread, are also those where garden types are less water-demanding. On the contrary, the larger municipalities of the compact metropolitan core are those where gardens with high water consumption are more commonly present. As we have already noted, public gardening is located basically in the largest municipalities of the compact city. Therefore and despite the fact that most municipalities, especially those of the second periphery, have low water-consumption gardens, in absolute terms, in the MRB high water-consumption gardens are dominant.

Most of the water consumed in the public spaces of the MRB is directed to the irrigation of turf grasses, since trees and shrubs are much smaller in area and the species used have moderate consumption rates. Thus, among the most common shrubs in public gardens of the MRB we found *Pitosporum tobira* or *Nerium oleander*, whereas, regarding trees, *Platanus x hibrida*, *Pinus pinaster*, *Pinus halepensis*, and *Olea europea* abound. All of the latter do not have excessive water requirements. In water sustainability terms, the main impact is that turf grasses not suited to local pluviometric regimes occupy more than half of the public-gardened area in the region.

In order to reduce water use—and, also in keeping with the objective of influencing choices in the private garden—there should be a change in the use of turf grasses. In this sense, there are indigenous species that could accomplish the same functions, especially when the main purpose is landscaping. If for certain reasons turf grass appears as necessary, then grasses more resistant to water stress, and therefore less water-consuming (*ovina, arundinacea, rubra, penisetum clandestinum* or *poa pratensis*, among others; Burés, 1993), could be used.

The key element in Xeriscaping, and crucial for environmental sustainability, is the type of garden. There are, however, other elements to be taken into account in garden sustainability practices such as mulching, the presence of dunes, the design of hidrozones (zones where species planted have similar water requirements), the destination of the redundant water, the irrigation system, or consumption control.

Xeriscaping is not common in the MRB despite the fact that some incipient initiatives are being implemented. The main limiting factor is the garden type, since the large presence of turf grasses makes Xeriscaping very difficult and also impedes the use of certain water-saving techniques. Sprinkler irrigation (albeit not the most efficient) and irrigation programming are common, but other management practices strike us as ultimately unsustainable. Among these, we must cite the lack of control in the use of irrigation (gardens remain unmetered) even when it rains and the noncollection of redundant water, which ends up in the sewers.

**TABLE 6.2. Categories of Species Planted
in the Gardened Areas of Public Spaces in the MRB**

Species planted	Percentage of gardened area	
	Compact city	Dispersed city
Trees	7	45
Shrubs	15	12
Turf grass	72	38
Other	6	5

Beyond the different techniques employed in Xeriscaping, the key issue in designing a garden based on sustainable criteria lies in selecting species well adapted to a Mediterranean environment. Otherwise, conservation practices lose relevance when applied to species with high water requirements and not adapted at all to climatic constraints. Hence, if most gardened areas are covered by high-consumption species, water conservation becomes virtually impossible.

From an environmental point of view, then, public gardening in the MRB lags behind standards in issue, such as careful species selection and Xeroscaping. The great paradox in environmental sustainability, however, is that public spaces located in low-density areas (where a large number of impacts are concentrated) are more sustainable than their counterparts in the compact city, mostly because they tend to use local species (see Table 6.2).

Social Sustainability: Public Spaces as Spaces of Relations

Public spaces are spaces lived in and modified in time through use. They constitute areas full of symbolism and significance in which society becomes visible (Lefebvre, 1974). Undoubtedly, public spaces nurture social relationships as long as they are public and therefore accessible to all. Moreover, relationships may take place outside the realm of privacy, and these relations reflect the society that inspires them. Even though there are other spaces where social relations can be developed, urban public spaces have three main properties that make them different from other potentially relational spaces.

The first property is free accessibility. As Borja and Muxí (2001) argue, free accessibility to public spaces turns them into democratic arenas or equal settings to which everybody is granted access. Public spaces are then spaces of citizenship to the extent that everybody may use them under conditions of equality. In turn, this stimulates diversity and tolerance toward differences.

The second property of urban public spaces is their openness. Urban public spaces are not only accessible but also "outside," in the open. The combination of both properties converts public spaces into meeting places and spaces to be used. Therefore, a certain public area may be simply a space for the circulation of people but also a space for meeting people. Finally, they may also be spaces for individual or collective uses with different purposes.

Being an open space allows public space to become much more than a space to go to on purpose to undertake any activity. It is a circulation space and hence a space of spontaneity. Other spaces that stimulate relational uses lack this property because they are utilized for certain specific functions that need individuals to make a purposeful decision about going there, either because they are closed spaces, open spaces but with restricted access, or because they have very specific functions, usually those linked to consumption (Zukin, 1998).

The third property of public spaces regards the multiple functionalities they possess. These are spaces where many uses are undertaken and activities are performed, both individual and collective: hiking, resting, sitting down, relating to others, playing, reading, celebrating, protesting, courting, enjoying, etc. Many of these activities are activities that encourage social relations, although it is also true that many of these functions can also be fulfilled individually. Public spaces allow for social relationships among people that know one another and stimulate these to the extent that these spaces are not only meeting areas but also spontaneous places for social relations, that is, places where two or more people can meet casually and initiate conversations. At the same time, public spaces are also places where new relationships may arise spontaneously among people not acquainted before (see Figure 6.5).

In the low-density areas of the MRB, public spaces, however, are not only residual but also very little used and therefore unable to fulfill their social functions. This assertion has been empirically contrasted through the observations made for this study in 19 public spaces of the MRB. While public spaces of the compact city were used by 100 or more persons at the same time, public spaces of the dispersed city (with similar dimensions and at the same time and date) were typically used by only 2 or 3 people. In many cases, we saw nobody during the 60-minute observation time period.

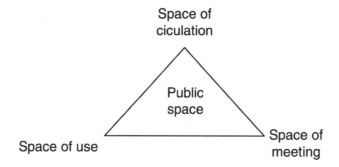

FIGURE 6.5. Properties of public spaces.

In the compact city, besides being relatively abundant, public spaces are used by a large diversity of people and collective bodies, during different times of the day, and with a thick density of users and relations. They are spaces that create citizenship that cultivates social interaction and cohesion, spontaneous meetings, and new relationships and that increase interpersonal trust. All together, public spaces contribute to the creation of social capital. Compact cities are diverse, and it is in the public spaces where this diversity becomes most visible, interacts, and generates democracy.

Recent territorial transformations, however, have led toward a segregation pattern that is not only functional but also social. The residential areas in the low-density city tend to be more socially homogeneous. With this, it is not just that there is a lack of public spaces to induce social interactions, but the heterogeneity necessary for these interactions is also missing. In the streets of these residential areas there is no social life, since they are usually empty and relationships become impossible. Dispersed urbanization has reduced public spaces to the benefit of private spaces at the same time that the lifestyles of residents have turned toward privacy.

Moreover, as an open and leisure space, public areas are not needed by residents of the dispersed city, since they already have their own private space. In parallel, the public space as space of relation loses relevance. The following comment extracted from the interviews illustrates this argument:

"All municipalities need public spaces, but, a municipality like this one, where all the houses are single-family houses with their own

garden . . . maybe it is not necessary to have a lot of public spaces."
(young woman, member of a voluntary organization, from a low-
income municipality and dispersed urban form)

The scarce public space of dispersed urbanization is, moreover, underused because residents follow a lifestyle oriented toward privacy. There
is yet another difficulty: most public spaces in these areas are far from
many of the residences.

In summary, public spaces in general are relational spaces with a
large potential to create social capital. Nevertheless, recent spatial
changes in the MRB have produced the proliferation of a low-density
model more prone to segregation, in which public spaces become residual and underused. If the opportunity to generate social cohesion is
missed, then this model can hardly be sustainable from a social point of
view.

Political Sustainability: From Public Spaces to Participation

From a global view of public spaces in the MRB, the analysis of the political dimension must not focus on whether these spaces have been initiated and developed from participative forms of government, since it is
quite clear that this has not been the case except on very few occasions.
What makes more sense is to observe whether these spaces may support
the creation, development, and success of new, more participatory, forms
of government that are the locus of political sustainability.

From our interviews with citizens, we have detected four aspects
through which public spaces may contribute to galvanize the development of new forms of government for complex societies based on the
active involvement of the citizenry. Thus, public spaces are, first, relational spaces in which people interact and speak about public affairs.
Second, these spaces create citizenship, since they stimulate relationships, cohesion, and identity. Third, they have been traditionally used
to voice political expressions of consent and dissent, and, fourth, they
are spaces that help generate a civic engagement of users with the
community. These four characteristics facilitate the development of
new government forms based on active participation by citizens (see
Figure 6.6).

To the extent that these are places of dialogue about public affairs,
public spaces stimulate debate about these affairs and therefore engage
citizens in all that is public:

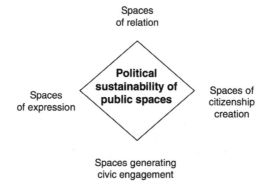

FIGURE 6.6. The components of the political sustainability of public spaces.

"In public spaces we talk about our city, about politics. . . . With people that I don't know much, I talk more about public affairs. Maybe with friends that are very close to me I talk more about private affairs." (adult woman, member of a voluntary organization, from a low-income municipality and compact urban form)

At the same time, public spaces are places of citizenry or spaces where the members of a community interact under the condition of free and equal citizens. The "exercise of citizenship" contributes to the involvement of each individual with his or her community, and this creates positive scenarios for the development of new governance forms.

As spaces of expression, public spaces are directly spaces of demonstration and participation where the citizenry may freely show its political options and demands. In one sense, we could also argue that public spaces are places where the people control the body politic. Therefore, their very existence and their democratic and free use are basic to the guarantees of any democratic form of government, and more so if the goal is to achieve more public participation. Finally, the generation of civic engagement is fundamental to the success of any participatory form of government. Without the involvement of the citizenry with the community, the collective construction of political action within a society makes little sense.

However, the critical element that best explains the potential of public spaces as spaces that, for different reasons, may prompt the success of new forms of governance is their relational character. This relational character is what allows for the existence of political conversations about the public affairs of any given community. The relational character inherent to public spaces also creates clusters of citizenship, since citizenship must be put into practice and requires places where this function can be accomplished. Citizenship, however, is not only performed individually but also collectively: the possibility of relationships with fellow citizens is also offered by public spaces because of their relational nature. In a similar manner, the generation of civic engagement arises also from the relational character of public spaces, because it is through relationships with others that citizens engage in the development of their societies (Ferris, Norman, & Sempik, 2001).

Summarizing, public spaces have a political dimension, because they are, most of the time, intrinsically social. But being political means more than being just social, because it fulfills some important aspirations such as the public expression of citizens' opinions, the demand and voicing of whatever is important for the citizenry, and the control of government (Mitchell, 1992, 1995).

Public participation, therefore, can benefit enormously from these spaces. However, in the low-density urban areas where, as we have seen, public spaces are residual, underused, and lacking social heterogeneity, this potential disappears. In the dispersed city, the arguments developed by Sennet (2002) prevail. This author affirms that the public sphere, as a political and social ideal, has become devalued and that the interest of the citizenry in public spaces has diminished. Despite the merit of these arguments, we would agree with others (e.g., Zukin, 1997; Ethington, 1994) that in the compact city, public spaces may turn highly diverse groups looking for social recognition and expressing particular demands into a more cohesive force acting in behalf of the wider community.

CONCLUSIONS

After our attempt to integrate scalar and transversal sustainability issues through this case study of public spaces in the Metropolitan Region of Barcelona, we can conclude that the current social model and its spatial manifestations have conditioned the creation of public spaces in the city, with a greater leverage for the compact areas and more privatization for the dispersed city.

As social *lieux* (place) for interaction, public spaces may breed the development of social capital and act as catalysts of emerging participative processes in our current society. That said, we have also noted how in our study area, while the need for new participatory democracy is arising, urban sprawl reduces the presence of public spaces and, in consequence, devalues an asset that may play a role in the constitution of emergent participative processes.

Thus, we are confronted with a paradox: one of the characteristics of the second modernity period envisaged by Beck, Giddens, and others is precisely the need for more participatory governance. At the same time, however, the spatial dilution of citizens in suburbia and the new urban lifestyles linked to urban sprawl may jeopardize the success of these participative endeavors, since the low-density form is deprived of public spaces and stimulates privacy, to the detriment of social capital.

From the standpoint of the transversal integration of sustainability, we have noted how in environmental terms—using water as a reference— the gardened areas of public spaces are characterized by a high consumption of this critical resource for Mediterranean environments. Despite their very low contribution to the total water spending of the MRB, if sustainable practices were more obviously present in the public gardens of the region, they could become a model for citizens to follow in their private gardens. The latter are increasing as a consequence of the proliferation of single and semidetached homes in our study area.

Water consumption in public gardening in the low-density areas is barely significant in environmental terms, since there are few public spaces and the plant species present require comparatively little water. However, in focusing on the social and political dimensions of sustainability, the impacts within low-density areas may be far larger than those of the compact city. A public space that, if present at all, is located amid scattered suburban homes remains unable to generate social capital, social identity, and civic cohesion. In this context, public spaces are neither social nor political. They are not social because they lack the capacity to foster social relations, and they are not political because they are not spaces of expression and civic compromise and cannot fuel the inputs necessary for the involvement of citizens in collective affairs.

In summary, the dispersed model of urbanization questions the concept of cities as places of diversity and places that create citizenship and calls into question public spaces as spaces that encourage social relations and political activism. Therefore, it is necessary to rethink the urban model and the role of public spaces from both a social and political sustainability logic. The social and political potentialities of public spaces are central to a broader conception of integrated sustainability and, in the context of the MRB, largely compensate for their relatively poor environmental performance. On the other hand, the higher environmental sustainability reached by public spaces in the low-density areas cannot offset their profound shortcomings in social and political terms. From the standpoint of scalar integration, the low-density model is deprived of social relations and, overall, is highly unsustainable.

In the MRB, it is possible to maintain the social and political character of urban public spaces and reduce environmental impacts, especially in water consumption, if plant species better adapted to the climate are used and Xeriscaping practices become more widespread. As Roseland (2000) would argue, this would be a perfect example of sustainable development in that social capital would increase and the consumption of natural capital would be reduced.

NOTE

1. The public sphere can also influence private gardening through incentives, norms, etc.

REFERENCES

Adams, W. (1990). *Green development: Environmental and sustainability in the third world*. London: Routledge.

Adger, N., Brown, K., Fairbrass, J., Jordan, A., Paavola, J., Rosendo, S., & Seyfang, G. (2003). Governance for sustainability: Towards a "thick" analysis of environmental decisionmaking. *Environment and Planning A, 35*, 1095–1110.

Albet, A., & Riera, P. (1998). Organització territorial i transformació urbana: De la ciutat industrial a la ciutat espectacle. Una entrevista amb Francesco Indovina. *Documents d'Anàlisi Geogràfica, 33*, 109–117.

Badia, A. (2000). *La incidència dels incendis a l'Àrea Metropolitana de Barcelona i a la comarca del Bages durant el període 1987–1998*. Mimeo, Universitat Autònoma de Barcelona.

Barr, S. (2003). Strategies for sustainability: Citizens and responsible environmental behaviour. *Area, 35*, 227–240.

Bauman, Z. (1992). *Intimations of postmodernity*. London: Routledge.

Bauman, Z. (2001). *Globalització: Les conseqüències humanes*. Barcelona: Edicions de la UOC.

Beck, U. (1998). *La sociedad del riesgo*. Barcelona: Paidós.

Beck, U. (2002). *Libertad o capitalismo*. Barcelona: Paidós.

Blanco, I., & Gomà, R. (2002). Proximidad y participación: Marco conceptual y presentación de experiencias. In I. Blanco & R. Gomà (Eds.), *Gobiernos locales y redes participativas*. Barcelona: Ariel.

Borja, J., & Muxí, Z. (2001). *L'espai públic: Ciutat i ciutadania*. Barcelona: Diputació de Barcelona.

Bridge, G., & Watson, S. (2000). *A companion to the city*. Oxford, UK: Blackwell.

Brugué, Q., & Gomà, R. (1998). *Gobiernos locales y políticas públicas*. Barcelona: Ariel.

Burés, S. (1993). *Xerojardineria*. Reus, Spain: Ediciones de Horticultura.

Burés, S. (2000). *Avances en xerojardinería*. Reus, Spain: Ediciones de Horticultura.

Burgess, J., Harrison, C., & Filius, P. (1998). Environmental communication and the cultural politics of environmental citizenship. *Environment and Planning A, 30*, 1445–1460.

Burton, E. (2001, April). *The compact city and social justice: Housing, environment and sustainability*. Paper presented at the Housing Studies Spring Conference, University of York, York, UK.

Domene, E., & Saurí, D. (2003). Modelos urbanos y consumo de agua: El riego de jardines privados en la Región Metropolitana de Barcelona. *Investigaciones Geográficas, 32*, 5–18.

Ethington, P. (1994). *The public city*. Cambridge, UK: Cambridge University Press.

Fenster, T. (2004). *The global city and the Holy city: Narratives on planning, knowledge and diversity*. London: Pearson.

Ferris, J., Norman, C., & Sempik, J. (2001). People, land and sustainability: Community gardens and the social dimension of Sustainable Development. *Social and Policy Administration, 35*, 559–568.

Garbancho, P. (1995). *La conquesta del verd*. Barcelona: Ajuntament de Barcelona.

Garcés, J., Ródenas, F., & Sanjosé, V. (2003). Towards a new welfare state: The social sustainability principle and health care strategies. *Health Policy, 65*, 201–215.

Giddens, A. (1999). *Consecuencias de la modernidad*. Madrid: Alianza.

Giddens, A. (2000). *Un mundo desbocado: Los efectos de la globalización en nuestras vidas*. Madrid: Taurus.

Gowdy, J., & O'Hara, S. (1997). Weak sustainability and viable technologies. *Ecological Economics, 22*, 239–247.

Harvey, D. (2003). *Espacios de esperanza*. Madrid: Akal.

Hediger, W. (2000). Sustainable development and social welfare. *Ecological Economics, 32*, 481–492.

Hirsch, F. (1980). *Social limits to growth*. Cambridge, MA: Harvard University Press.

Jameson, F. (1991). *El Postmodernismo o la lógica cultural del capitalismo avanzado*. Barcelona: Paidós.

Kates, R., Clark, W., Corell, R., Hall, J., Jaeger, C., & Lowe, I. (2001). Sustainability science. *Science, 292*, 641–642.

Lefebvre, H. (1974). *La production de l'espace*. Paris: Anthropos.

Macnaghten, P., & Jacobs, M. (1997). Public identification with sustainable development: Investigating cultural barriers to participation. *Global Environmental Change, 7*, 5–24.

Marne, P. (2001). Whose public space was it anyway?: Class, gender and ethnicity in the creation of the Sefton and Stanley Parks, Liverpool: 1858–1872. *Social and Cultural Geography, 2*, 421–443.

Mas-Collell, A. (1994). Elogio del crecimiento económico. In J. Nadal (Ed.), *El mundo que viene*. Madrid: Alianza.

Massey, D. (1994). *Space, place and gender*. Oxford, UK: Blackwell.

Mitchell, D. (1992). Iconography and locational conflict from the underside: Free speech, People's Park, and the politics of homelessness in Berkeley, California. *Political Geography, 11*, 152–169.

Mitchell, D. (1995). The end of public space? People's Park, definitions of public, and democracy. *Annals of the Association of American Geographers, 85*, 108–133.

Muñoz, F. (2004). *Urbanalització: La producció residencial de baixa densitat a la província de Barcelona, 1985–2001*. Unpublished papaer, Universitat Autònoma de Barcelona.

Nel·lo, O. (2001). *Ciutat de ciutats*. Barcelona: Editorial Empúries.

Pain, R. (2000). Place, social relations and the fear of crime: A review. *Progress in Human Geography, 24*, 365–387.

Parés, M., Domene, E., & Saurí, D. (2004). Gestión del agua en la jardinería pública y privada de la Región Metropolitana de Barcelona. *Boletín de la Asociación de Geógrafos Españoles, 37*, 223–237.

Putnam, R., Leonardi, R., & Nanetti, R. (1994). *Making democracy work: Civic traditions in modern Italy*. Princeton, NJ: Princeton University Press.

Ravetz, J. (2000). Integrated assessment for sustainability appraisal in cities and regions. *Environmental Impact Assessment Review, 20*, 31–64.

Robbins, P. (2004). *Political ecology: A critical introduction*. London: Blackwell.

Roseland, M. (2000). Sustainable community development: Integrating environmental, economic, and social objectives. *Progress in Planning, 54*, 73–132.

Rotmans, J., Asselt, M. V., & Vellinga, P. (2000). An integrated planning tool for sustainable cities. *Environmental Impact Assessment Review, 20*, 265–276.

Rueda, S. (1995). *Ecologia Urbana: Barcelona i la seva regió metropolitana com a referents*. Barcelona: Beta.

Sassen, S. (1998). *Globalization and its discontents. Selected essays*. New York: New York Press.

Saurí, D. (2003). Lights and shadows of urban water demand management: The case of the Metropolitan Region of Barcelona. *European Planning Studies, 33*, 230–243.

Saurí, D., Capellades, M., Rivera, M., & Paredes, A. (2001). An analysis of the Residential Water Sector in the Metropolitan Region of Barcelona. In P. Firma (Ed.), *Bellaterra*. Barcelona: Universitat Autònoma de Barcelona (unpublished).

Sennet, R. (2002). *El declive del hombre público*. Barcelona: Edicons 62.

Spangenberg, J., & Bonniot, O. (1998). *Sustainability Indicators: A compass on the road towards sustainibility*. Unpublished manuscript, Wuppertal Institute.

Speir, C. & , Stephenson, K. (2002). Does sprawl cost us all? Isolating the effects of housing patterns on public water and sewer costs. *Journal of the American Planning Association, 68*, 56–70.

Stocker, G. (1996). Governance as theory: Five propositions. *International Social Sciences Journal, 155*, 17–28.

Subirats, J. (1997). Democràcia: participació i eficiència. *Revista CIFA, 6*, 4–17.

Swyngedouw, E. (1997). Neither global nor local: Glocalisation and the politics of scale. In K. Cox (Ed.), *Spaces of globalization: Reasserting the power of the local*. London: Longman.

Tello, E. (2000). *La dinàmica socioecològica del Baix Maresme als anys noranta: L'onada residencial i els seus impactes ambientals i socials*. Unpublished manuscript, Mataró.

Tjallingii, S. (2000). Ecology on the edge. *Landscape and Urban Planning, 48*, 103–119.

Valentin, A., & Spangenberg, J. (2000). A guide to community sustainability indicators. *Environmental Impact Assessment Review, 20*, 381–392.

Zukin, S. (1997). *The cultures of cities*. Oxford, UK: Blackwell.

Zukin, S. (1998). Urban lifestyles: Diversity and standardisation in spaces of consumption. *Urban Studies, 35*, 825–839.

CHAPTER 7

Political Modernization and the Weakening of Sustainable Development in Britain

ANNA BATCHELOR
ALAN PATTERSON

This chapter discusses the significance of the structural framework within which policies and practices for sustainable development are being advanced at the local and regional levels in the United Kingdom. Hams (1994) argued that local authorities have a real and growing interest in sustainable development, and both Hams and Levett (1994) emphasize the need for policy integration and a strong corporate approach by local authorities in addressing these issues. However, it is also important to consider how far British local authorities can progress toward sustainable development, given their current functions and powers.

Community involvement and participation are seen as crucial to the environmental policy process, with the need to involve all sectors of the community; as the "Bruntland Report" argued: "Sustainable development requires a political system that secures effective citizen participation in decision making" (World Commission on the Environment and Development, 1987: 65). There also appears to be general agreement that, to fulfill the requirements of sustainable development, there is a need for democratic and holistic local and regional authorities and for the integration of sustainability into mainstream policies and practices (Agyeman & Evans, 1994; Tuxworth, 1994; Carter, 2001). It is clear, therefore, that to initiate successful policies for sustainable development subnational government institutions need to reassess their role and engage directly with their constituent communities. Moreover, the impact of such involvement depends upon such authorities having the power, competence, and resources to act upon the results of consultations with community groups. Sustainable development requires both subsidiarity and democratization to permit participation and empowerment at the subnational level. This chapter therefore raises questions about the ability of British local and regional authorities to initiate and enact sustainable development policies.

Pattie and Hall (1994) argue that there are significant barriers to implementing environmental strategies at the local level, including the complexity and interrelatedness of environmental issues, the lack of adequate resources, and the fact that most local government activity is defined by narrow statutory responsibilities that restrict capacity for discretionary action. Our research shows that, at the structural level too, the long-term processes of reorganizing subnational governance have negative implications for local authorities seeking to develop positive and inclusive environmental strategies, and the latest form of restructuring— founded on the principles of "political modernization"—has yet further reduced their capacity to achieve meaningful local solutions to the problems of sustainable development. "Political modernization" is, however, only the latest in a long-running series of changes in the structure, functions, and powers of local authorities that have removed former local government functions to a variety of nonelected agencies and central departments (Patterson & Pinch, 1995).

Since the election of the Labour government in 1997 subnational government within the United Kingdom has undergone another round of fundamental reforms. The creation of a Scottish Parliament, Welsh Assembly, and Greater London Authority, along with a program of local government modernization and strengthening of the regions, has resulted in constitutional changes. It would be overly simplistic to portray this program as wholly negative: the Welsh Assembly was established with a statutory duty to promote sustainable development in all of its policies and spending programs, and the new English regional institutions have been charged with a variety of policy responsibilities relating to sustainable development. However, it is a premise of this chapter that understanding the changing powers, functions, and struc-

tural arrangements of subnational government is crucial to understanding why the policy framework for sustainable development is
weakening.

Raco (2005: 327) describes a theoretical tension between those who
believe the state is engaged in rolling out neoliberalism and those who
believe it is pursuing sustainable development (SD): "For at the same
time as the principles of SD have come to 'dominate' policy agendas,
others argue that it is neoliberalism, with its principles of market efficiencies, entrepreneurial communities, and resource exploitation which
have, paradoxically, taken centre stage." In this chapter we aim to show
how this tension is being played out in the subnational government
structures in the United Kingdom. There is certainly a strong rhetoric of
sustainable development—and this dominates policy agendas and debates—
but the structures of governance and policy determination continue to
ensure that, in practice, there remains a strong focus on the implementation of (neoliberal) policies for continued economic development and the
weak and "muddled" implementation of policies for sustainable development.

This complexity is addressed by drawing on empirical case studies
of local and regional governments in the South East of England.

However, before examining the implications of the most recent
(post-1997) round of restructuring, it will be useful to recap the changes
that have already taken place.

RESTRUCTURING LOCAL GOVERNMENT

One hundred years ago local government in Britain was in its heyday, as
Burgess and Travers (1980: 21) put it: at the end of the 19th century "local government was responsible for most of the activities of government
apart from Defence." However, as the decades passed, more and more
powers and functions were removed from the local level and transferred
to central departments, unelected *ad hoc* bodies, or to the private sector.
One of the earliest indicators of the trend was the government's insis-
tence in 1905 that London's water supply should not be controlled by
the London County Council, an elected body with a "radical" reputation, but should instead be controlled by the Metropolitan Water Board,
a body specially created for the purpose and run by a board of nominees. However, for others the year 1934, when responsibility for poor

relief was transferred from local authorities to the national Unemployment Assistance Board, marked the beginning of the decline of local government autonomy (Dearlove & Saunders, 2000: 306).

Later the responsibility for property valuation (a task that had been conducted at the local level for 350 years) was transferred to the Inland Revenue, heralding the introduction of a new form of equalization grant for local authorities that ushered in the current era of close central control of local government expenditure. The courts and the auditors have also acted to restrict the authority and autonomy of local authorities through the use of the concepts of *"ultra vires"* and "fiduciary duty," for example, to outlaw a policy of low public transport fares in London in 1986. As Hams (1994: 205) argued, local authorities have been "quangoed to death. Everything's about accountancy rather than accountability." Although local authorities gained some responsibilities during this period—particularly in relation to planning and social services—compared to the situation at the beginning of the 20th century, local authorities are now much larger and more bureaucratic and impersonal; they perform fewer functions and have less autonomy over those that they have retained; and their scope for imaginative policy implementation is tightly constrained both financially and legally.

This "hollowing out" of local government (Patterson & Pinch, 1995) continued with the introduction of compulsory competitive tendering (CCT) for the provision of many local services (Pinch & Patterson, 2000). CCT is an important case because it led to the fragmentation of local authority responsibilities as individual departments lost direct provision of services to the private sector. Rather than encouraging holistic local authority decision making, CCT required the separation of budgets and responsibilities. As Levett (1994) and Hoyles (1994) note, local authorities are major players in the local economy and the local environment, and therefore they have a powerful impact on the environment through the implementation of their policies. However, the requirement, under the CCT legislation, to accept the lowest offered bid reduced the power of the local state to act in support of the local economy by purchasing goods and services from local firms through local purchasing arrangements that could provide a contribution to sustainable development. Furthermore, CCT specifically excluded the consideration of the activities of a contractor in terms of the stance it adopted on economic, social, and environmental issues (see Patterson & Theobald, 1995, 1996, 1999).

POLITICAL MODERNIZATION

Since the election of the Labour government in 1997, local government has undergone a further series of fundamental alterations through a program of change known as political modernization. This has been characterized by a shift from the clear distinction between the state and external bodies, to the rise of quasi-governmental bodies and more collaboration with external organizations (Rydin, 1999). In opposition Labour had pledged to abolish CCT; however, instead it introduced Best Value—requiring local authorities to subject more areas of provision to competition from the private sector and to undertake service reviews and publish service and performance plans, all of which were externally inspected (Ball, Broadbent, & Moore, 2002; Downe & Martin, 2006). With the Best Value legislation maintaining a strong emphasis on market-driven decision making, the privatization or quasiprivatization of public services, and extending the influence of the market into many new areas (Andrews, 2003; Geddes & Martin, 2000), the process of "hollowing out" local government has not been reversed.

Modernization seeks to alter local government's role, with a new emphasis on partnerships among local authorities, business, and the community/voluntary sector. This is what Giddens (1998) has termed the "third way," which he sees as a rejection of both the (left-wing) interventionist role for the state *and* the (right-wing) opposition to state involvement, instead putting forward an agenda for governance through partnership. According to Giddens, this "third way politics looks for a new relationship between the individual and the community, a redefinition of rights and obligations" (1998: 65). Tony Blair (1998) has highlighted how central this "third-way" program of modernization is for New Labour's plans for local government, stating that a renewal of local democracy is required specifically in order to tackle social exclusion and to implement Local Agenda 21 (LA21).

While the modernization program is upheld as a new way forward for local government, Hill (2000) argues that New Labour's program follows that of the "enabling" state approach of the former Conservative administration. Hill notes that the government has argued that its partnership approach to governance will bring greater skills and expertise to local government from the private sector as opposed to the Conservatives' approach of forcing a business culture onto local authorities in an attempt to undermine its power. However, the partnership approach can come into conflict with democratic accountability; thus, Hill (2000: 7) states: "the question to be addressed is whether we are entering a new era of a 'third way' between state collectivist solutions and *laissez faire* capitalism, as Labour claims, or just seeing the stabilisation of a public–private provision of services that have emerged from the previous Conservative revolution."

Four Examples of Political Modernization in Practice

Reform of the Local Government Committee System

One of the most far-reaching of the modernization changes has been the change to the decision-making structure of local government. Required to move from the well-established committee system, local authorities had to choose from one of three options:

- A directly elected mayor with cabinet—the elected mayor to appoint the cabinet members from among the councilors.
- A cabinet with a leader—the leader to be elected by councilors, and the cabinet either elected by councilors or appointed by the leader from among the councilors.
- A directly elected mayor and council manager—the mayor to provide political leadership to the council manager but not to make day-to-day decisions.

The justification for this change was the claim that the old system was inefficient and lacked transparency. However, it has led to the concentration of power and the creation of a two-tier system of councilors, resulting in what the government review team (Stoker, Gains, Greasley, John, & Rao, 2004) observed as the great dissatisfaction of nonexecutive councilors with the new arrangements, and it does not appear to have led to a discernible pattern of improved political leadership (Leach, Hartley, Lowndes, Wilson, & Downes, 2005).

Local Strategic Partnerships and Community Strategies

Local strategic partnerships (LSPs) should bring together representatives from public, private, and community sectors with the objective of joining up disparate programs and initiatives (Department for Environment,

Transport, and the Regions [DETR], 2001a). LSPs are nonstatutory (but the government recommends that all local authorities establish one), and their core task is to create the statutory Community Strategies that are to act as the new overarching framework for public, private, and community sector activities within the local area. Community Strategies are supposed to incorporate the aims of sustainable development, and therefore the government has stated that preexisting LA21 strategies should be subsumed within them (DETR, 2001b). More recently the government has promoted a shift to Sustainable Community Strategies, which are intended to: "evolve from Community Strategies to give a greater emphasis to sustainable development objectives" (Department for Environment, Food and Rural Affairs [DEFRA], 2005: 127). However, the primary focus here relates to the creation of sustainable *communities* "which are necessary for creating an area where people genuinely want to live long-term" (DEFRA, 2005: 127) and is far removed from the original strong environmental focus of LA21, which has now been superseded.

The guidance for the LSP and Sustainable Community Strategies espoused the involvement of all parties, but there has been disquiet among elected members, with many councilors we interviewed feeling that their role was being undermined and the democratic functions of local government were being eroded.

"Well-Being"

The new power to promote "well-being" permits local authorities to undertake activities that promote or improve the economic, social, or environmental well-being of the area. This is a discretionary power of general competence that can be used to do anything that other legislation does not expressly forbid, but it does not enable local government to raise money for these undertakings. When utilizing the well-being power, local authorities must take account of their Community Strategies. In theory at least, the power of well-being offers a proactive role to local government in promoting the interests, and enhancing the welfare, of its community. However, with understanding of the power described as "patchy" and its use generally confined to discretionary rather than mainstream services, the power is deemed by local authorities to be weak and to only partly address the doctrine of *ultra vires* (Sullivan et al., 2006).

In practice the use of the well-being power is tightly constrained by limits on expenditure and concern over potential litigation that may arise from its use. In the past, the use of powers derived from the Local Government Act 1972 (which allowed local authorities to undertake actions deemed to be in the best interest of the community) often resulted in legal challenges, and the courts often adopted a narrow interpretation of the legislation. Therefore, although the well-being power appears to offer an opportunity to local government to promote sustainable development, legal and financial barriers remain.

Comprehensive Performance Assessment

Comprehensive Performance Assessment (CPA) is an inspection regime used to assess local authority performance against a set of centrally determined criteria. Inspections are conducted annually by the Audit Commission and currently result in a rating of one to four stars and a "direction of travel" that indicates whether a local authority's performance is improving or worsening (Audit Commission, 2005). This rating is used by the government to allocate extra funding and greater autonomy, what Downe and Martin (2006) describe as "earned autonomy." Therefore, CPA has become a priority for local authorities. CPA focuses on traditional local authority service areas and neglects crosscutting issues such as sustainable development, but also, because there is no explicit reference to sustainable development in the inspection criteria, CPA further marginalizes this policy area (Miller, 2002; Bennett, 2003).

The government has declared that it intends to change the standing of sustainable development within CPA inspections, stating that it "will seek to recognise and reward good performance on SD" (DEFRA, 2005: 161). However, it is questionable how far CPA, which is fundamentally designed to assess the efficiency of service delivery, can be developed to address the issue of sustainable development, particularly when this policy area is being marginalized in so many other ways.

The following section presents a brief case study of a local authority within the South East of England that had strongly embraced the environmental elements of sustainable development but which has shifted its focus in response to the modernization policies outlined above.

Reading Borough Council: A Case Study

With a long history of community development work, Reading Borough Council (RBC) was one of the first local authorities to respond positively to LA21 by developing the widely reported "go local on a better environment" (GLOBE) ward-based community environment groups (e.g., see Buckingham-Hatfield & Percy, 1999; Parker & Selman, 1999). GLOBE groups were designed to enable direct community participation in the council's environmental decision-making process, and a multidisciplinary team of officers was set up with two roles:

- internally—to get the local authority's own house in order, through the development of such processes as environmental management systems; and
- externally—to work with the community on themed projects.

These initiatives were specifically carried out under the LA21 banner, with a clear and explicit environmental mandate. Between 1993 and 1996 RBC developed its position on environmental issues, moving from an "Environmental Statement of Intent" to a more proactive policy approach through LA21, and created an "LA21 Team" that included community development workers and worked in new areas—such as establishing a farmers' market. Community involvement featured strongly in the development of RBC's LA21 strategy, with a focus on participation in local environmental problem solving and policymaking, empowered by the policy of Agenda 21, through initiatives such as "neighbourhood action plans" (Reading Borough Council, 2000).

During the 1990s "a shift in emphasis from 'local government and the environment' to one of 'local governance and sustainability' " (Parker & Selman, 1999: 18) occurred. In part this was due to the recognition of the inability to address "environmental" issues without recognizing the inextricable link to social and economic concerns. In Reading, this shift toward integrating LA21 into other initiatives ("mainstreaming") began with the Sustainable Communities Dialogue, which sought to engage with local groups to identify specific actions that could be taken to make Reading more sustainable. Themes derived from the dialogue formed the basis of RBC's local strategic partnership, which then became the primary mechanism for community representation. This change in approach directly reflected the shift in central government guidance which specified the need to embed LA21 into the Best Value regime, the LSPs, and the statutory Community Strategies. But, although mainstreaming LA21 issues could perhaps have resulted in a more integrated approach (e.g., RBC's three corporate strategic aims all now embed aspects of sustainability), it also had the effect of shifting the focus from environmental concerns to a more anthropocentric "quality-of-life" agenda (Batchelor & Patterson, 2004).

RBC's LSP has been developed through a representative process, and it has sustainable development as one of its key objectives, but the national guidance does not demand this. The inclusiveness of Reading's LSP has been achieved through the commitment of the local authority's officers, *despite* the weakness of the national legislation. Therefore, developing equitable community involvement in the new "modernization" initiatives nationally is a vulnerable and uneven process.

This vulnerability is particularly prevalent in relation to resources. The mainstreaming of sustainable development within Reading was a major factor in the recent disbanding of the "LA21 Team" and its replacement by a smaller sustainable development team with a corporate focus, based in the chief executive's office. In a recent spending review the new team's budget was cut severely, as it was not seen to provide either a core or a statutory service.

ENVIRONMENTAL MODERNIZATION

Jacobs (1999) argues that the U.K. government's unwillingness to put environmental concerns at the heart of policy is based on the belief that environmental concerns are inextricably linked to a green ideology that is viewed as fundamentally anticapitalist, and therefore, as New Labour does not share this view, the concept of sustainable development has not been embraced. Jacobs believes that environmental concerns can be separated from green ideology in a way that mirrors New Labour's transition to the "third way," and which, therefore, could permit environmental issues to become part of mainstream New Labour politics. Giddens (using the term "ecological modernisation") agrees, seeing a potentially more cohesive path for New Labour politics and the environment, because "there is no doubt that ecological modernisation links social democratic and ecological concerns more closely than once seemed possible" (1998: 57).

Dryzek believes a third-way alliance can be formed "in which governments, businesses, moderate environmentalists and scientists co-operate in the restricting of the capitalist political economy along more environmentally defensible lines" (1997: 145). However, as Blowers (1999) highlights, the concept of ecological or environmental modernization has become prominent precisely because it proposes that the objectives of economic growth and the environment are *not* in conflict, which firmly locates it within New Labour's contemporary political discourse of political modernization and partnership. Therefore, environmental modernization can be seen as portraying a "business as usual" scenario (Hajer, 1995).

Beck's (1992) theory of the Risk Society presents a challenge to this cosy congruence, as it places the concept of "ecological risk" at the center of policy concerns. As Blowers (1994: 14) comments, "ecological modernisation provides the case for the continuation of an environmentally sensitive form of modernisation; risk society confronts the necessity for change." While Giddens (1998) assumes groups such as NGOs can be assimilated within the third-way agenda, Beck (1992) argues that politically uniting forces such as class have disintegrated, leading to what he describes as "sub-politics," which represents a *challenge* to conventional democratic political decision-making forums through the pursuit of single-issue politics by pressure groups.

This brief discussion has attempted to highlight some of the significant differences between the environmental movement and the discourse of New Labour in relation to environmental concerns. While sustainable development may have been developed by the "pragmatic wing" of the green movement (Jacobs, 1999) in order to put environmental concerns into mainstream politics, this has certainly not received the support of the government. LA21 has not been supported strategically by central government, and most of the innovative work that has taken place has been undertaken by relatively small groups of committed local authority officers and community activists, and, as illustrated in Reading, many of these successes have been relatively short-lived.

The government's political modernization program may be able to be used to facilitate some of the aims of sustainable development, but, as Evans and Percy (1999) noted in the case of LA21, they may also just produce another round of consultations and fail to achieve community empowerment. Moreover, new initiatives that have been introduced, such as "Best Value," have a strong neoliberal orientation and have hindered rather than helped promote sustainable development.

Regional Government in the South East of England: A Case Study

Often referred to as the "growth engine" of the United Kingdom's economy, the South East (SE) of England provides a useful regional scale case study to examine the tensions in balancing the pursuit of economic growth against the desire for sustainability. With a population of over 8 million and an economy generating over £140 billion per year (which makes a £20 billion net annual contribution to the Exchequer) the SE is the largest of the United Kingdom's regions (South East England Regional Assembly [SEERA], 2004a). The region's economic preeminence is associated with its geographical location, particularly its proximity to London and mainland Europe, and good transport connections, particularly the Channel tunnel and Heathrow airport.

Although the existence of regions for administrative purposes has a long history in the United Kingdom, the current structure was determined in 1992 when England was divided into nine regions for the purpose of creating the Government Offices (GOs). The GOs brought together the regional interests of several central government departments and were established to enhance regional-level collaboration with the European Union and to facilitate the implementation of central government policy at the regional scale (Allen, 2001). As the Government Office for the South East (GOSE) states: "We represent central government in the region and our role is to promote better and more effective integration of Government policies and programmes at a regional and local level" (Government Office for the South East, 2006). Since 1998, Regional Development Agencies (RDAs) and Regional Assemblies (RAs) have joined the GOs as institutions of governance at the regional level. Within the SE these are the South East England Development Agency (SEEDA) and the South East England Regional Assembly (SEERA). The RDAs' purpose is to promote economic development, enhance business, employment, and the job skills base, and contribute to sustainable development. SEEDA states that it is "responsible for the sustainable economic development and regeneration of the South East of England" (SEEDA, 2006). One way it seeks to implement this responsibility is through its 10-year Regional Economic Strategy (RES) (SEEDA, 2002), which was developed in conjunction with GOSE and SEERA.

SEERA (2004b) describes itself as the "voluntary regional chamber" for the SE ("voluntary" because, although assemblies have been created in all of the regions, the government did not make their creation obligatory), with three core functions:

- *Accountability*—specifically scrutinizing the work of SEEDA.
- *Advocacy*—acting as the "voice of the region" to influence national and EU agendas.
- *Planning*—in 2001 SEERA became the Regional Planning Body, and this activity now accounts for 70% of its work and resources.

SEERA has 112 members, none of whom is directly elected, including 74 nominees from the constituent local authorities, and others representing business interests and the voluntary/community sector. Initially the Regional Assemblies were funded entirely by the local authorities and other regional interests, but since 2001 they have also received some funding from the central government (Office of the Deputy Prime Minister, 2003).

SEERA has taken the lead on two overarching regional strategies—the Integrated Regional Framework (IRF) and the Regional Spatial Strategy (RSS)—and must be consulted by SEEDA during the production of its RES (ODPM, 2006). Through the IRF SEERA (2004c) aims to establish "a shared vision and objectives for integrated working and ultimately, sustainable development of the region." The RSS covers a 20-year period and is intended to set the context for local land-use planning and transport strategies: "Core objectives are to balance continuing economic and housing growth with rising standards of environmental management and reduced levels of social exclusion and natural resource consumption. Our vision for 2026 is for a healthier region, a more sustainable pattern of development and a dynamic and robust economy, the benefits of which are more widely shared" (SEERA, 2006: 2).

GOSE, SEEDA, and SEERA have developed cooperative working practices and are beginning to foster what Musson and Tickell (2005) describe as a "regional political culture." However, in practice, aligning policies is challenging, because there are different priorities and, as explored below, issues of overlapping objectives. While the IRF may seek to provide a coherent shared vision for the region, it does not have statutory status and so is relatively weak in comparison to the RSS and the RES. Although the IRF is developed by regional institutions and the RSS and RES emerge from central government policy (albeit with some regional input), there is some ambiguity about the spatial scale at which the regional agenda is being set: "Government Offices work with regional partners to develop, implement and monitor 'Regional Spatial Strategies,' which set out *Government's* planning and transport policy for each region for a 15–20 year period" (Government Office of the South East, 2006a, emphasis added).

A new tier of government is clearly being developed at the regional scale, but, as the previous quotation highlights, it is not an autonomous level and it is difficult to see exactly where the power lies. However, a recent review noted: "the very prominent role played by SEEDA in regional strategy making" (SEERA, 2004c). The objectives and work programs of both SEEDA and GOSE have been formed by the central government, and although the GOs were created by a Conservative government, it is now New Labour ideology that shapes these institutions: "It was clear that RDAs encapsulated the 'new' in New Labour" (Musson & Tickell, 2005: 1400). Moreover, as the SE region is the "engine room" of the United Kingdom's economy, it is difficult to envisage any specific regional issue being allowed to override national policies. For example, as the Sustainable Communities Plan for the South East states: "We cannot simply try to halt growth in the South East in order to divert it to other regions. The government's regional policy is focused on enabling every region of England to perform to its full economic and employment opportunities" (Office of the Deputy Prime Minister, 2003: 5). Although a regional policy agenda is developing, the region is clearly not autonomous; however, at the same time, GOSE and SEEDA are honing their policies in line with regionally defined objectives. Therefore, the governance of the SE is becoming increasingly complex.

Policies developed at the regional level are made in partnership. SEERA notes that this ensures compliance across the regional bodies, stating that "we have worked closely with the RDA in a number of key policy areas to ensure complementarity, such as housing, sustainable development, renewable energy, waste markets, transport and urban renaissance" (SEERA, 2004b: 11). With SEERA and SEEDA each having responsibilities for sustainable development and GOSE also having such interests, policy on this topic is diffuse. This dispersal of responsibility leads to a lack of clear leadership and creates what Hewett (2001) has described as "institutional muddle." To address these issues the following section examines sustainable development policy and practice in the region and assesses New Labour's approach to the issue.

"INSTITUTIONAL MUDDLE": REGIONAL BODIES
AND SUSTAINABLE DEVELOPMENT

The creation of a uniform regional approach in the SE, under the umbrella of the Integrated Regional Framework, has proved to be complex. As noted above, SEERA, the regional assembly, took the lead in producing the IRF, an overarching regional document that specifies the involvement of other regional bodies and requires conformity of their policies with respect to sustainable development. On this basis SEERA might appear to be the lead body for sustainable development at the regional level. However, SEERA, like all regional assemblies, only has the responsibility to *promote* sustainable development, and most of this work was undertaken *after* the SEEDA had developed the Regional Economic Strategy (Hewett, 2001). This weakness of the regional assembly's role is evident in the SE, with SEEDA taking the lead on the move toward "Smart Growth,"[1] which has become the latest incarnation for sustainable development in the SE.

One of the five aims on which RDAs were established was a duty "to contribute to sustainable development" (DETR, 1998). As Figure 7.1 illustrates, SEEDA asserts that its RES is "set within the broader context of sustainable development" (South East England Development Agency, 2002: 8).

With the aim of managing what has often been described as an "over-heated economy" that is suffering from the problems of economic success (e.g. Foley, 2004), SEEDA has adopted the concept of "smart growth" in an attempt to justify continued economic growth. SEEDA (2003: 2) argues that "smart growth" could be achieved through a 75% reduction in the SE's ecological footprint, to be attained by "doubling resource efficiency and halving resource usage," and sets out key indicators that would be monitored, arguing that increased productivity can be consistent with sustainable development.

The concept of "smart growth" represents a paradigm shift in policymaking for sustainable development. The concept came into use in the United States during the 1990s in relation to urban regeneration (Krueger, 2005). Noting its successful incorporation into the policies of local and regional agencies, compared with the limited success of the adoption of LA21, Krueger notes: "In the U.S. 'smart growth' has emerged as an American variant to the Bruntland paradigm of SD" (2005: 78). To understand why "smart growth" is being so readily adopted in the South East of England, we need only revisit the concept

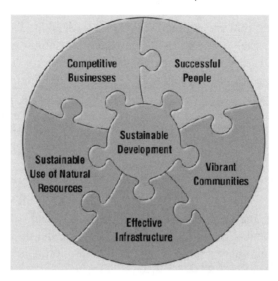

FIGURE 7.1. The five objectives of the Regional Economic Strategy. Data from SEEDA (2002).

of environmental modernization. "Smart growth" allows a neoliberal "business-as-usual" scenario, because, as Krueger (2005) emphasizes, at its core it is an economic development strategy. It fits well with the aims and objectives of SEEDA and the guiding principle on which all regional assemblies were founded: the promotion of economic growth.

While the Government Offices do not have explicit responsibilities for sustainable development, GOSE states that SD is embedded within its priorities (Government Office of the South East, 2006), and within the South East region the main involvement of the GO has been through the development of the Regional Spatial Strategy. Arguing that the GOs are the most influential of the regional institutions, Musson and Tickell (2005) highlight their role in providing expert advice on policy development. They state that this has the effect of projecting central government aims into the regions. If this is considered in relation to the production of the RSS, GOSE's role could have a significant effect upon SD policy in the region.

With the three institutions each have differing roles and responsibilities toward sustainable development, and an overlap of objectives, it is hard to ascertain where the lead is and on which ideology this is based. However, to achieve an objective as broad as sustainable development requires policy coordination and clear leadership, but the institutional muddle ensures that the regional scale lacks leadership. This was one of the key issues that participants in the recent consultation for the U.K. Sustainable Development Strategy highlighted, with six regions recommending that a statutory obligation for sustainable development be established, because there was: "strong support for putting the delivery of sustainable development at the regional level on a statutory footing" (DEFRA, 2004). Despite this call, the government states only that it will apply a statutory duty to any new bodies created (although such a duty already exists for the National Assembly for Wales and the Greater London Assembly) and that it will "assess whether a specific SD duty should be applied to existing key bodies in priority areas" (DEFRA, 2005: 156–157). Instead, the strategy outlines proposals for each of the regional bodies: Regional Assemblies' role in sustainable development will be clarified through new guidance, sustainable development will be mainstreamed through the RDAs' "Tasking Frameworks," and GOs will be subject to new performance and monitoring arrangements. Perhaps the most significant development is the enhanced role for the Sustainable Development Commission (SDC), as a "watchdog" on sustainable development.

The government (DEFRA, 2005: 152) specifically notes the importance of "those providing public services at regional and local level" in achieving sustainable development, but falls short of explicitly giving statutory responsibility to local or regional authorities. To understand why, we need to examine the ideology of New Labour. By deferring a decision, the government is afforded the opportunity to see what works and what effects SD policies have at the regional scale before committing to institutional reform—exactly what Stoker (2002) describes as a "purposeful muddle." Moreover, as discussed earlier, New Labour's environmental ideology aligns with the principles of environmental modernization. Worried about the consequences of embracing environmental concerns, that ideology is carefully placing these concerns within the sphere of institutions created by and working within neoliberal agendas. The ability to develop counterhegemonic coalitions is thereby limited by state structures (Gibbs, Jonas, & While, 2002: 133).

CONCLUSIONS

The continuing restructuring of subnational government institutions has been heavily influenced by the ideologies of political and environmental modernization. Fundamentally changing the nature of local government and creating an unelected regional tier has resulted in the weakening of sustainable development and the implementation of neoliberal policies such as "smart growth."

As the Reading case study illustrates, the government's decision to subsume LA21 strategies into the modernization initiatives and to mainstream sustainable development policy into the new "quality-of-life" agenda has resulted in a sidelining of the core *environmental* aspects of sustainable development and, with it, the core participatory aims of LA21. Councils have, understandably, shifted their efforts to focus on the new performance regimes, such as the rigorous Comprehensive Performance Assessment. As Porritt (2000) has noted: "Local authorities were given powers to promote the 'economic, social and environmental well-being' of local people, but the guidance on community planning and local strategic partnerships that followed was written as if central government had never heard of sustainable development."

Prior to the introduction of LA21, local government in Britain had been denuded of powers and functions, and under the Conservative Government CCT and other market-driven policies further reduced local authorities' capacity to respond to the demands of sustainable development. Although New Labour promised much, including the abolition of CCT, the policies it introduced (e.g., Best Value) did not increase local governments' ability to achieve sustainable development, because the heavy emphasis on neoliberal solutions and private sector provision were retained. Although many local authorities demonstrated a willingness to respond positively to LA21, the restructuring of their functions and powers reduced their capacity to respond adequately; and the contemporary program of "political modernization" is not resolving these problems—rather, it is further sidelining sustainable development as a policy goal.

Along with the modernization program for local government has come the creation of a new regional tier of governance that is shaping policy for sustainable development. This is resulting in the region becoming a significant administrative tier—but without clear leadership. The regional level is impacting upon the autonomy and power of local government, which, despite having a democratic mandate, must now operate in conformity with the regional agenda; and the new national inspection regimes (e.g., Best Value and CPA) are giving central government more authority over local authorities. Concepts such as Stoker's (2002) "purposeful muddle" have sought to explain the rationale behind the government's program of change. Geddes and Martin (2000) note that the program was purposefully vague, to enable the government to test what would work, and Downe and Martin (2006) see this in a positive light, as an "evolutionary strategy."

The 2005 U.K. Sustainable Development Strategy has set out the government's position on priorities for sustainable development (including an initiative, Community 2020, which looks remarkably like a New Labour variant of LA21). However, the strengthening of neoliberal structures and the constrained capacities of subnational government will continue to inhibit the development of meaningful policies for sustainable development in Britain.

NOTE

1. This chapter refers to "smart growth" in the British policy context, which stems from the theory of "Factor Four," developed by von Weizsäcker, Lovins, and Lovins (1997). SEEDA (2003) adopted this concept in its "Taking Stock" report, which reviewed resource usage in the SE. This report posits that a 75% reduction in the ecological footprint of the SE can be achieved by doubling resource efficiency while halving resource use—the key idea of Factor Four. The term "smart growth" as used within this context relates to the decoupling of resource use from economic growth.

REFERENCES

Agyeman, J., & Evans, B. (1994, July/August). Making Local Agenda 21 work: *Town and Country Planning, 91*, 197–198.

Allen, A. (2001). Overview of institutions. In C. Hewett (Ed.), *Sustainable development and the English regions*. London: Institute for Public Policy Research and the Green Alliance.

Andrews, G. (2003, Winter). Modernisation or marketisation? *Soundings, 22*, 7–10.

Audit Commission. (2005). *CPA—the harder test framework for 2006*. London: Audit Commission.

Ball, A., Broadbent, J., & Moore, C. (2002, April–June). Best Value and the control of local government: Challenges and contradictions. *Public Money and Management, 22*, 9–16.

Batchelor, A., & Patterson, A. (2004, March). *Institutional and organizational obstacles to local sustainable development*. Paper presented at annual meeting of the Association of American Geographers, Philadelphia.

Beck, U. (1992). *The risk society*. London: Sage.

Bennett, J. (2003, February). Travelling into unchartered waters: The experience so far of integrating sustainability into CPA. *eg magazine, 9*, 7–9.

Blair, T. (1998). *The Third Way*. London: The Fabian Society.

Blowers, A. (1999, June). *Ecological and political modernisation: The challenge for planning*. Paper presented at the local economic regeneration—social and environmental impacts conference. Brunel University, Uxbridge, UK.

Buckingham-Hatfield, S., & Percy, S. (1999). Keys to a sustainable environment: Education, community development and local democracy. In S. Buckingham-Hatfield & S. Percy (Eds.), *Constructing local environmental agendas: People, places and participation* (pp. 1–17). London: Routledge.

Burgess, T., & Travers, T. (1980). *Ten billion pounds*. London: McIntyre.

Carter, N. (2001). *The politics of the environment: Ideas, activism, policy.* Cambridge, UK: Cambridge University Press

Dearlove, J., & Saunders, P. (2000). *Introduction to British politics* (3rd ed.). Cambridge, UK: Polity Press.

Department for Environment Food and Rural Affairs. (2004). *Taking it on: Developing UK sustainable strategy together*. London: DEFRA.

Department for Environment Food and Rural Affairs. (2005). *Securing the future: UK Government sustainable development strategy*. London: The Stationery Office.

Department for Environment, Transport, and the Regions. (1998). *Modern Local Government: In touch with the people*. London: Her Majesty's Stationery Office.

Department for Environment, Transport, and the Regions. (2001a). *Local Strategic Partnerships: Guidance*. London: Her Majesty's Stationery Office.

Department for Environment, Transport, and the Regions. (2001b). *Strong Local Leadership, Quality Public Services*. London: Her Majesty's Stationery Office.

Downe, J., & Martin, S. (2006). Joined up policy in practice?: The coherence and impacts of the Local Government Modernisation Agenda. *Local Government Studies, 32*(4), 465–488.

Dryzek, J. (1997). *The politics of the earth*. Oxford, UK: Oxford University Press.

Evans, B., & Percy, S. (1999). The opportunities and challenges for local environmental policy and action in the UK. In S. Buckingham-Hatfield & S. Percy (Eds.), *Constructing local environmental agendas: People, places and participation* (pp. 172–185). London: Routledge.

Foley, J. (2004). *The problems of success: Reconciling economic growth and quality of life in the South East*. London: Institute for Public Policy Research.

Geddes, M., & Martin, S. (2000). The policy and politics of best value: Currents, cross-currents and undercurrents in the new regime. *Policy and Politics, 28*, 379–395.

Gibbs, D., Jonas, A., & While, A. (2002). Changing governance structures and the environment: Economy–environment relations at the local and regional scales. *Journal of Environmental Policy and Planning, 4*, 123–138.

Giddens, A. (1998). *The third way: The renewal of social democracy*. London: Polity Press.

Government Office for the South East. (2006a). Accessed August 20, 2006, at *www.gose.gov.uk*.

Government Office for the South East. (2006b). Available at *www.gose.gov.uk/gose/environmentRural/sustainableDev/?a=42496*.

Hajer, M. (1995). *The politics of environmental discourse: Environmental modernisation and the policy process*. Oxford, UK: Oxford University Press.

Hams, T. (1994, July/August). International agenda, local initiative. *Town and Country Planning, 91*, 204–205.

Hewett, C. (2001). *Sustainable development and the English regions*. London: Institute for Public Policy Research and the Green Alliance.

Hill, D. M. (2000). *Urban policy and politics in Britain*. London: Macmillan.

Hoyles, M. (1994). *Lost connections and new directions: The private garden and the public park* (The Future of Urban Parks and Open Spaces Working Paper No. 6).

Jacobs, M. (1999). *Environmental modernisation: The New Labour agenda*. London: The Fabian Society.

Krueger, R. (2005). Governing Smart Growth: A critical examination of the American variant of sustainable development. In R. Krueger (Ed.), *Proceedings of the Regional Studies Association Annual Conference: Sustainable regions—making regions work* (pp. 78–79). London: Regional Studies Association.

Leach, S., Hartley, J., Lowndes, V., Wilson, D., & Downe, J. (2005). *Local political leadership in England and Wales*. London : Joseph Rowntree Foundation.

Levett, R. (1994, July/August). Options from a menu. *Town and Country Planning, 91*, 206–207.

Miller, A. (2002, August/September). Riding the storm: CPA and sustainable development. *eg magazine, 8*, 13–16.

Musson, S., & Tickell, A. (2005). A decade of decentralisation? Assessing the role of the Government Offices for the English Regions. *Environment and Planning A, 37*, 1395–1412.

Office of the Deputy Prime Minister. (2003). *Sustainable Communities Plan for the South East*. London: Her Majesty's Stationery Office.

Parker, J., & Selman, P. (1999). Local government, local people and Local Agenda 21. In S. Buckingham-Hatfield & S. Percy (Eds.), *Constructing local environmental agendas: People, places and participation*. London: Routledge.

Patterson, A., & Pinch, P. L. (1995). Hollowing out the local state: Compulsory competitive tendering and the restructuring of British public sector services. *Environment and Planning A, 27*(9), 1437–1461.

Patterson, A., & Theobald, K. S. (1995). Sustainable development, Agenda 21, and the new local governance in Britain. *Regional Studies, 29*(8), 773–778.

Patterson, A., & Theobald, K. S. (1996). Local Agenda 21, compulsory competitive tendering and local environmental practices. *Local Environment, 1*(1), 7–19.

Patterson, A., & Theobald, K. S. (1999). Emerging contradictions: Sustainable development and the new local governance. In S. Buckingham-Hatfield & S. Percy (Eds.), *Constructing local environmental agendas: People, places and participation* (pp. 156–171). London: Routledge.

Pattie, K., & Hall, G. (1994). The greening of local government: A survey. *Local Government Studies, 20*(3), 458–485.

Pinch, P. L., & Patterson, A. (2000). Public sector restructuring and regional development: The impact of compulsory competitive tendering in the UK. *Regional Studies, 34*(3), 265–275.

Porritt, J. (2002, October 1). Can she sustain it? *The Guardian*. Available at *http://www.guardian.co.uk/comment/story/o,,802094,00.html*.

Raco, M. (2005). Sustainable development, rolled-out neoliberalism and sustainable communities. *Antipode, 37*, 324–347.

Reading Borough Council. (2000). *Reading's Local Agenda 21*. Reading, UK: Author.

Rydin, Y. (1999). Environmental governance for sustainable urban development: A European model? *Local Environment, 4*(1), 61–66.

South East England Development Agency. (2002). *Regional economic strategy for South East England 2002–2012*. Guildford, UK: Author.

South East England Development Agency. (2003). *Taking stock: Managing our impact—an ecological footprint of the South East region.* Guildford, UK: Author.

South East England Development Agency. (2006). Accessed August 20, 2006, at *www.seeda.co.uk.*

South East England Regional Assembly. (2004a). *Fact sheet.* Guildford, UK: Author.

South East England Regional Assembly. (2004b). *Frequently asked questions.* Guildford, UK: Author.

South East England Regional Assembly. (2004c). CAG review of key regional strategies in the South East. In SEERA, *Integrated regional framework compendium.* Guildford, UK: Author.

South East England Regional Assembly. (2006). Regional Assembly grant support. Accessed March 8, 2006, at *www.odpm.gov.uk/index.asp?id=1109130.*

Stoker, G. (2002). Life is a lottery: New Labour's strategy for the reform of devolved governance. *Public Administration, 80*(3), 417–434.

Stoker, G., Gains, F., Greasley, S., John, P., & Rao, N. (ELG Evaluation Team). (2004). *Operating the New Council Constitutions in English Local Authorities: A process evaluation.* London: Office of the Deputy Prime Minister.

Sullivan, H., Rogers, S., Crawford, C., Kitchin, H., Evans, L., & Mathur, N. (2006). *Formative evaluation of the take-up and implementation of the well being power.* London: Department of Communities and Local Government.

Tuxworth, B. (1994, July/August). Blazing a Local Agenda 21 trail. *Town and Country Planning, 91,* 212–213.

von Weizsäcker, E. U., Lovins, A. B., & Lovins, L. H. (1997). *Factor four: Doubling wealth, halving resource use.* London: Earthscan.

World Commission on the Environment and Development. (1987). *Our common future.* Oxford, UK: Oxford University Press.

CHAPTER 8

Spatial Policy, Sustainability, and State Restructuring

A Reassessment of Sustainable Community Building in England

MIKE RACO

When the Blair government was elected in 1997, it committed itself to the modernization of the British planning system and to the encouragement of new types of community-driven, partnership-based local governance (see Imrie & Raco, 2003). Through a series of new legislation and policy strategies it sought to change the modus operandi of spatial planning and redefine its core priorities and objectives. During the 2000s these reforms have been increasingly subsumed within the wider parallel discourse of sustainable development that has also become a guiding principle for governments across the developed world. These new ways of thinking were reflected in the publication in 2003 of the Blair government's most significant statement on the foundations of English spatial policy, the *Sustainable Communities Plan*, in which the planning system would be retooled to deliver places "where people want to live and work now and in the future. They meet the diverse needs of existing and future residents, are sensitive to their environment, and contribute to a high quality of life" (Office of the Deputy Prime Minister, 2005: 1).

The emergence of these sustainable communities (SCs), as the new objects and subjects of spatial policy, has raised a series of questions concerning broader processes of state regulation and socioeconomic and environmental change. Where, for example, have ideas about SCs, as understood in the English context, come from? Why have they come to the fore in spatial policy discourses at this particular moment? What does their emergence tell us about the broader relationships between places and economic competitiveness and the rationalities and philosophies that underpin contemporary spatial policy? Moreover, in what ways do they reflect a broader concern with environmental limits to development and questions of social and spatial justice?

This chapter explores and develops these questions through an assessment of the early phases of the SC agenda in England. The main argument will be that during its first 3 years it has represented anything but a coherent and logical spatial development program. It has been fraught with implementational difficulties that have reflected and reproduced the incoherence of its core aims and objectives. In an earlier article (see Raco, 2005) I argued that the program had to be understood as a hybrid, appearing somewhere on a continuum between what might be termed "neoliberal" and "sustainability" development philosophies. In this chapter I suggest that even this paints an unrealistic picture of its coherence and that, in reality, the agenda has been undermined by the limited *capacities* of the British state to deliver on its rhetorical objectives. The program has also been met with challenges at the national, regional, and local scales and has been (re)shaped less by wider ideologies about the appropriate role for spatial policy and more by the practical need to overcome implementation problems in specific development areas. These findings, it is argued, have conceptual and methodological implications for the ways in which spatial policy and (welfare) state restructuring are characterized.

This chapter explores these issues through four interrelated sections. The first and second document the rise of the SC agenda and outline its key aims, characteristics, and objectives. They situate the new agendas in the context of wider policy debates over sustainability and community-driven urban development. This is followed by a third section that assesses the relationships between the new agendas and what they reveal about broader processes of state regulation. It examines the extent to which the proposals reflect and reproduce a wider shift toward more selective and divisive post-Keynesian spatial policies and explores the role of the environment and of more grounded forms of politics in shaping the form and character of state regulation and activity. This is followed by a final section that examines the prospects and possibilities inherent in the SC program and outlines some of the key trends that are likely to have an impact on future spatial development agendas in the United Kingdom and beyond. It also highlights potential directions for future research.

THE EMERGENCE OF THE SUSTAINABLE COMMUNITY

In February 2003 the Labour government published its *Sustainable Communities Plan* that promised to create a "step change" in the planning process and to enable SCs to be planned and constructed. The plan sets out a vision for new-build settlements in the South East of England and the regeneration of urban centers across the United Kingdom. Community building is to be targeted in three growth areas in the South East—Milton Keynes, the Cambridge–Stansted airport corridor, and Ashford, where at least 260,000 new homes are to be built. In addition, a minimum of 120,000 homes have been proposed for an area known as the Thames Gateway that stretches along the Thames from East London to North Kent and South Essex, an area that first became identified as a development zone during the 1980s, when local industries and docks closed at the same time that property markets in the west of London were overheating and suffering from growing supply-side restrictions (Heseltine, 1990; Lawless, 1989). Under the SC proposals this development area has been earmarked for massive long-term redevelopment in order to improve the "balance" of development between the west and east of the region and to provide space for further expansion of globally competitive industries and the labor markets required to service them. These developments have been supplemented by the creation of nine market renewal areas in northern cities in which there will be "sustained action to replace obsolete housing with modern sustainable accommodation, through demolition and new building or refurbishment" (Office of the Deputy Prime Minister, 2003: 24).

The core features of an SC are outlined in Table 8.1. They indicate that a sustainable place is one in which a "balance" of employment, mixed housing, and social facilities are copresent and available to a range of socioeconomic groups. In essence an SC is, therefore, an inherently spatial or geographical construct whose scale varies according to the specific policy contexts in which it is defined. Its construction in-

**TABLE 8.1. The Central Features of Sustainable
and Unsustainable Communities**

Criterion	A sustainable community	An unsustainable community
Economic growth	Flourishing economic base; built on long-term commitments; stable; and inclusive of a broad range of workers.	Domination by dependent forms of development; lack of employment opportunities; vulnerable; insecure, short-term; and divisive.
Citizenship	Active citizens and communities; long-term community stewardship; effective political engagement; healthy voluntary sector and strong social capital.	Passive and dependent citizens and communities; lack of community engagement or ownership; low levels of voluntary activity and/or social capital.
Governance	Representative, accountable governance systems; balance of strategic, top-down visionary politics and bottom-up emphasis on inclusion.	Closed, unaccountable systems of governance; overreliance on passive, representative forms of democracy; lack of visionary politics; parochialism.
Community characteristics	Broad range of skills within workforce; ethnically and socially diverse; mixture of socioeconomic types of inhabitants; balanced community; well-populated neighborhoods.	Absence of skills within workforce; ill-balanced communities of place; high levels of (physical) separation between groups; lack of diversity; formal and informal segregation; lack of population.
Urban design	Diverse architecture; accessible public spaces; higher urban densities; provision of a broad range of amenities; buildings that cater to a range of needs; "self-contained" communities; the creation of "place."	Uniform zoned architecture; closed, gated, and inaccessible public spaces; absence of community facilities; urban sprawl; "placeless" suburban development.
Environmental dimensions	Reuse of brownfield sites; minimization of transport journeys; high-quality public transport	Expansion into greenfield sites; maximization of transport journeys; car dependence and the absence of public transport.

(continued)

TABLE 8.1. (*continued*)

Criterion	A sustainable community	An unsustainable community
Quality of life	Attractive environments; a high quality of life; strong pull for a range of social groups.	Low quality of life; strong push for a range of social groups.
Identity, belonging, and safety	Sense of community identity and belonging; tolerance, respect, and engagement between people of diverse backgrounds; low levels of crime and antisocial behavior.	Lack of local associational culture and ownership of public space; intolerant and divided local politics; high levels of crime, disorder, and fear.

Note. Data from Raco (2007; adapted from Office of the Deputy Prime Minister, 2003, 2005).

volves the integration and copresence in time and space of a particular type of built environment, a diverse and broad range of employment and employees, and a degree of social cohesion that facilitates a well-functioning social order. The agenda is underpinned by particular visions of what places could and should be like in order to be balanced and sustainable and highlights the processes of mobility and fixity through which such places can be made and remade. It is not simply the movement of people that drives the new agendas but the selected mobilization of diverse social groups, the (spatial) shifting of housing development priorities, the movement of jobs, shifting priorities for infrastructure investment, and the creation of new types of leisure facilities. As such, the Labour government's understanding of an SC represents something of a hybrid between the principles of sustainable development, broadly defined, and a progrowth strategy for enhancing place economic competitiveness. The role of spatial planning becomes that of "fixing" imbalances and ordering space and place so that they become more functional, cohesive, and competitive.

Since 2003, the core elements of this SCs discourse have become guiding principles for the wider modernization of the English planning system. In 2004, for example, the government rewrote the guidelines that set out the fundamental principles of local and regional planning with its publication of *Planning Policy Statement 1: Delivering Sustainable Development* (Office of the Deputy Prime Minister, 2004). These guidelines make it a legal requirement that local and regional planners address a range of interconnected social, economic, and environmental issues when permitting any development to take place and also consider the ways in which their individual decisions contribute to the wider principles of SC building. Moreover, the guidelines also stress that effective spatial planning is essential to any SC program and that more broadly "a spatial planning approach should be at the heart of planning for sustainable development" (p. 6). In many ways these new guidelines represent the culmination of longer-term changes in policy thinking and practice that go beyond the programs of the Blair government, both temporally and spatially. They draw on two interrelated development objects—those of *sustainability* and *community*—that have become dominant during the 1990s and 2000s, each of which is briefly discussed in turn in the next section.

THE TWIN DISCOURSES
OF SUSTAINABILITY AND COMMUNITY

The discourse of *sustainability* has become one of the central guiding orthodoxies of planning across Europe and North America (De Roo & Miller, 2000). Planners now seek to create sustainable places, cities, and regions in which development can be supported by social, economic, and environmental resources in the long term for the benefit of individuals, communities, and society as a whole (see Gibbs, 2002; Pearce, Markandya, & Barbier, 1991). The term originally referred to new forms of ecological modernization in which it was envisaged that development processes could become more democratic, environmentally sustainable, and more equitable for existing and future generations (see Whitehead, 2003). In the words of James Meadowcroft, the diffusion of this way of thinking reflects a certain irony in that "just at a time when philosophers have proclaimed the death of 'meta-narratives' . . . international political leaders have come to identify themselves with an ambitious new project intended to act as the focus for human endeavour in the 21st century" (Meadowcroft, 1999: 12). The adoption of sustainability as a "meta-narrative," Meadowcroft argues, is a reflection of its fluid and context-dependent meanings and the ways in which it can be used to justify a range of policy programs, from the planning of "compact cities" to the creation of "mixed" communities and more environmentally friendly urban design (see Burton, 2000; Jenks, Burton, & Williams, 1996).

However, despite being labeled a "metadiscourse," sustainability's widespread acceptance and actually occurring forms in particular places cannot be simply read off from a wider global logic. Its fluid meanings have enabled different groups of actors to redefine and reconceptualize it in different ways for their own political ends (see Harvey, 1996). As Maloutas (2003) notes, sustainability only takes on specific meanings and substance through *processes of recontextualization* in different local, regional, and national environments. This recontextualization is a politically structured process in which sustainability can take on different meanings and forms that can even "result in it being used to support opposite points of view" (Counsell, 1999: 46).

At the same time, in the United Kingdom and other EU countries many spatial policy programs have been focused on the related concept of *community-led development*. Its advocates argue that state actors and experts no longer carry the legitimacy and authority to draw up and implement development programs. Instead, there is a new emphasis on what Giddens (2005) terms the "co-production" of public goods and policy initiatives in which there "should be collaboration between the state and the citizen in the production of socially desirable outcomes" (p. 16). This "co-production" involves a redrawing of the boundaries between modern states and citizens so that the *responsibility* for policy programs is shared, with the state's role limited to providing "certain guarantees that the state has a moral and political responsibility to provide" (Schuppert, 2005: 57). Communities, it is argued, can draw on stocks of local knowledge and expertise in ways that make development programs more efficient and effective. Successful community mobilization can also encourage individuals to become more active so that they take more responsibility for shaping their own lives and the character of the communities in which they live.

This new emphasis on government through community has been taken up with great zeal by the Blair administration in the United Kingdom and by the European Commission[1] as an important part of their broader strategies to modernize (and downscale) welfare state systems and establish new relationships between citizens and states (see Imrie & Raco, 2003; Rose, 1999). Responsibilizing communities in this way reduces the moral and political justification for a well-resourced redistributive welfare state system. Through a range of initiatives and programs policymakers are now expected to identify, mobilize, and activate communities in the pursuit of wider objectives. In the longer term, it is argued, this form of community activation will change the character of democratic systems of governance and move them away from more traditional representative systems of democracy and accountability toward more participative and deliberative forms of direct engagement.

The emergence of the *sustainable community* therefore represents something of a hybrid between these broader development discourses. It involves a logical extension of the wider discourse of sustainability when applied to a British context in which social and spatial inequalities are the highest in the EU and the principle of government through community has become a relatively well-established part of welfare state reform and efforts to tackle the negative impacts of this inequality.[2] In this sense SCs are ostensibly designed to promote and institutionalize the broader principles of sustainability planning while developing new types of active and inclusive forms of community building and citizenship. In a more optimistic sense they can be portrayed as a positive step toward more holistic ways of defining and understanding spatial policy problems and possibilities. At the same time, however, they can also be criticized for their lack of clarity and for being too broad-ranging and ill defined in scope. Whatever the discourse's relative strengths and weaknesses, its central role in contemporary spatial policy discourses also sheds light on wider processes of state regulation and the objectives and practices of policy, and it is to these themes that the chapter now turns.

THE SUSTAINABLE COMMUNITIES AGENDA AND NEW POLITICS OF SPACE?

The emergence of the SCs' agenda in the United Kingdom and elsewhere has taken place in a context where, many commentators now argue, spatial planning and policy have become dominated by new development philosophies. There are two principal dimensions of change, both of which relate to the new discourses of spatial policy that are being promoted, not only in the United Kingdom but across the EU. These are (1) a New Regionalism and (2) a shift toward more social consumption-based forms of spatial development policy. Each will now be examined in turn.

The Rise of the New Regionalism

For a number of authors, particularly those writing from a regulationist perspective, the imaginations and institutional systems that shape and implement spatial policy across the developed world have been subject to significant and ongoing change. The spaces of policy intervention have moved away from the preexisting Keynesian concern with the strength of the *national* economy to one focused on the identification, entrepreneurial mobilization, and competitiveness of selected *regions*. Under spatial Keynesianism policymakers were concerned with "equalizing the distribution of industry, population and infrastructure across national territories" through a conception of "the entire national economy as an integrated, auto-centric, self-enclosed territorial unit moving along a linear developmental trajectory" (Brenner, 2003: 207; see also Gamble, 1988; Harvey, 2005). In the United Kingdom the Distribution of Industry Act of 1945 established the development principles that would shape spatial policy until the late 1970s. It legislated for an ordered redistribution of population and industry, with the explicit aim of supporting the social and economic development of the poorest regions. The prevailing logic was that state power could and should be used to increase the efficiency of the United Kingdom's economy and spread the socioeconomic benefits of growth across the regions. In the contemporary words of *The Economist* (1945: 270), this reflected "a new attitude, in which the scheduled areas are no longer regarded as plague-spots, to be diagnosed by specialists and treated as something apart from the rest of the community."

The conventional wisdom (see Gordon & Buck, 2005) among policymakers and many academics is that with the breakdown of the Keynesian settlement in the mid-1970s the character of spatial selectivity, state strategies, and state projects has been transformed so that states now "strive to differentiate national political-economic space through a re-concentration of economic capacities into strategic urban and regional growth centres" (Brenner, 2003: 207; see also Peck & Tickell, 2002). These so-called *new regionalist* strategies are underpinned by very different imaginations and ideologies about how spatial economies operate, the policy mechanisms that are necessary for national and regional competitiveness, and the moral and political priorities of the state (see Lovering, 2006; Scott, 2005). In contrast to Keynesian programs, this new regionalism downplays the *interrelationships* between "competitive" and "uncompetitive" places and promotes "the re-concentration of industrial growth and infrastructure investment within strategic urban and regional economies" (Brenner, 2003: 208). Policy becomes focused on identifying, selecting, and supporting the most economically successful regions in order to sustain and enhance their global competitiveness (Jones, 1997). It draws on new imaginings of how global economies work and the increasingly limited capacities of national states to directly shape the scale and location of economic activity. The emphasis is on helping regions to help themselves become more competitive rather than intervening to support all the regions of the country. The net effect of this new regionalism is that growth is concentrated in an increasingly small number of competitive centers that are able to capture wider flows of global investment and resources. Enhanced socioeconomic polarization is its inevitable result.

Spatial Policy and Social Consumption

The emphasis of new regionalist writing is on how states look to boost the productive capacities of places and the firms and individuals within them. However, spatial policy discourses in the "faster-growing" regions of the United Kingdom, the United States, and other Western countries are also becoming increasingly concerned with the relationships between production and *social consumption*. As authors such as While, Jonas, and Gibbs (2004) and Scott (2005) show, the greatest threat to the continuity of growth in such areas is a lack of supply in the availability of hard consumption assets such as housing and transport infrastructure, and the erosion of softer social and welfare services such as education and health care. Costs of living are increasingly so rapidly that low and moderately paid workers are being priced out of local housing markets, and this is having a significant impact on the quality and availability of social and economic systems. In addition, there is often growing pressure on environmental resources in such areas and the physical capacities of places to sustain significant amounts of further development.

Spatial planning is, therefore, becoming increasingly concerned with the provision of infrastructure that can mollify or reduce these development pressures. In the South of England, for example, policy programs such as the Key Worker Living Programme have been introduced to tackle housing constraints for some public sector workers (see Raco, 2007). New proposals are also being developed for the further "liberalization" of the planning system to enable new rounds of construction and development to take place, so that shortages can be effectively tackled (see Barker, 2006). The old-fashioned support that was given to industrialists to invest in their production facilities and/or to move their location(s) has given way to a new emphasis on providing the environments in which new forms of economic activity can flourish and removing some of the identified environmental "limits" to growth. In Harvey's

(1994) terms, this new spatial planning is seeking to create new spatial fixes in which the needs of accumulation are supported by the creation of new (built) environments.

Implementing the Sustainable Communities Agenda: Development Rationalities, Implementation Deficits, and the New Politics of Development

Given that the SC agenda has been promoted as a "step change" in spatial policy, we might expect it to reflect and reproduce the wider trends in policy thinking and practice discussed above. A provisional reading of the new development blueprints appears to exemplify a new regionalist dimension and a consumption-based orientation to policy. There is, for example, little or no discussion of the ways in which state agencies, through the planning system, can directly intervene to regulate and control the economic geographies of the United Kingdom. There is an implicit recognition that the state's role, in this new global age, is to ensure that thriving industries and regions sustain their success and that their economic growth underpins national social cohesion and capital accumulation. Rather than focusing on the direct ordering of national economic spaces, a new regionalist imagination is presented in which a core objective is explicitly to "accommodate the economic success of London and the wider South East [of England] and ensure that the international competitiveness of the region is sustained, for the benefit of the region and the whole country" (Office of the Deputy Prime Minister, 2003: 46). The underlying rationality is that "we cannot simply try to halt growth in the South East in order to divert it to other regions" (p. 5). Instead, the SC plan is described as being part of a wider, new regionalist policy shift that is "focused on enabling every region of England to perform to its full economic and employment opportunities" (p. 5; see also Raco, 2005).

In addition, the SC plans also incorporate new ways of thinking about the relationships between economic production and social and environmental consumption. In seeking to "accommodate" economic success, the emphasis is increasingly on the planning of infrastructure and the skewing of social spending toward "competitiveness ends." The construction of SCs requires the copresence of skilled workers, entrepreneurs, and public sector key workers, and this, in turn, requires significant investments in former industrial, brownfield land and a major expansion in the supply of affordable housing. It is this emphasis on using the spatial planning system to create environments that are conducive to new forms of economic development and community building that represents the most explicit recognition that consumption pressures are limiting the future sustainability of growth and development and that policy should be used to tackle these limits to growth.

However, the new discourses, representations, and imaginations that underpin the SC agenda represent only a part of the wider picture of policy change. Imaginations and forms of discursive representation are, of course, critical in shaping policy agendas. The "rolling back" of more interventionist spatial policy in the early 1980s was premised upon a sustained and politically powerful critique of "big government" (see Peck & Tickell, 2002). And yet, the emergence of the SCs agenda has to be conceptualized in terms of both their rationalities *and* their grounded social, environmental, political, and economic impacts. There can be a significant difference between the aspirations of policymakers and the institutional structures and resources that exist, or are created, to bring policy measures to fruition. Implementation deficits, for example, are a recurring feature of policy-making processes in developed countries, and it is only through an assessment of how and why policy programs are developed and delivered that we can gain a wider insight into their effectiveness and significance (see Hogwood & Gunn, 1984, for a broader discussion). Indeed, it could be argued that in the broader field of urban and regional studies there has been too much of a shift away from concrete analysis and too much attention paid to the construction of discursive imaginations and representations (see for example Amin & Thrift, 2002).

The SC agenda provides an excellent example of this distinction between discourses and implementation practices. Since its launch in 2003, the U.K. government has provided neither the resources nor the new institutional powers required by local and regional planning agencies to implement the SC proposals in the ways envisaged by its policy discourses. In contrast to the state-led agendas of postwar spatial policy, the mechanisms through which employers, citizens, and communities will be "encouraged" to relocate to the new areas are unclear and limited (for a discussion of the history of spatial policy in the United Kingdom, see Ward, 2004). New types of mobility and fixity will be encouraged through public sector-led investments in existing brownfield sites to make them more attractive, but there is little in the way of direct government grants or payments to firms or individuals. The plan, for example, provides only £448 million for new investment in the Thames

Gateway area—an area of 81,000 hectares, possessing 38,000 hectares of previously used brownfield sites (Campaign for the Protection of Rural England, 2004). Such a sum is utterly inadequate to meet the wider and grander visions for SC building in the area. Its private sector-led vision of development is unlikely to be successful, given the existing problems within the house-building sector in the United Kingdom, with, for example, household growth outpacing housing units growth by 59,000 per year between 1999 and 2004. The government's Barker Review of housing in 2004 showed that 14,000 fewer houses were built in London and the South East between 1996 and 2001 than the government's target figure (Barker, 2004). House-building rates at this time were at their lowest since 1924 at a time when economic growth and demographic changes were pushing up demand for a range of housing, particularly in London and the South East of England.

One of the ironies of the SC agenda, therefore, is that in seeking to deliver a "step change" in the planning process and the refashioning of spatial policies to support the expansion of faster-growing regions, the Labour government has run up against the organizational limits of the British state system. It has found that building SCs will require more than the development of new visions and discourses. In the absence of strong redistributive welfare state structures, the resources and powers needed to fulfill its wider targets and ambitions currently do not exist, and there is little provision for them in the new frameworks. In their place, a new voluntarism is being promoted in which individuals, investors, and firms make their own investment and (re)location decisions, with government's role becoming one of gentle encouragement and aspiration building, thus reflecting a longer set of traditions within the British welfare state system (see Houlihan, 1988). In order for states to be more supportive of selected fast-growing regions, as new regionalists claim (see Brenner, 2003), they may ironically have to increase their capacities and become more interventionist. Indeed, it is possible that increasing the scale of state involvement may generate new types of urban and regional policies and/or lead to new forms of neo-Keynesian redistributive politics and policies.

In addition to these organizational weaknesses, the implementation of any spatial policy agenda is bound up with the broader politics of spatial development and the process of *governance*. The emergence of the SCs agenda and its implementation to date have been dependent on a grounded interregional politics of spatial development taking place across different spatial scales. For example, during the 2000s, the strongest calls for enhanced government intervention in the spatial economy have come, ironically, not from the regions with the greatest socioeconomic problems but from those in which the pressures of development have been at their strongest. Thus, the Mayor of London, Ken Livingstone, has been at the forefront of debates over spatial policy and has sought to mobilize a consensus politics in the capital in which calls are made for the central state to "give back" the tax revenues that the city's taxpayers pay into the national Treasury. In his provocative words, "[London] generates more wealth than any other region in the country, contributes more to national finances, and makes a unique contribution to the nation's prosperity. Sustaining London's progress has to be a national priority" (Livingstone, 2004: 3).

Other agencies, such as London First and the South East England Development Agency (2003), have also been promoting the idea that the needs of competitive places have to be met by the state in order to guarantee their longer-term survival. The SC plans are as much a reflection of the Labour government's attempts to mollify and respond to this grounded and contested politics as they are of abstract and ideological imaginations about state capacities, globalization, and the role of British regional and urban economies within them.

However, it is other forms of grounded politics concerning environmental and quality-of-life issues that have had, arguably, the most significant impact on the SC agenda during the first 3 years of its existence. One of the biggest impediments to implementation has been the physical and environmental limits that have been placed on the building of new communities and the regeneration of existing ones. Government quangos (quasi-nongovernmental organizations) such as the Environment Agency (2005) and English Nature (renamed Natural England in 2007) (2006) have argued that new development projects in London and the South East will be unable to overcome the severe environmental pressures and limits that are already affecting the region, particularly in relation to water supplies and the quality of open space (British Broadcasting Corporation, 2006). Even on brownfield sites, where the government has promoted redevelopment as an environmentally sustainable process, there are growing criticisms that the new agendas will have a detrimental effect on the diversity and vibrancy of urban plant and animal species. Urban brownfields often provide significant refuges for wildlife, and their destruction or regeneration can undermine broader efforts to protect endangered species.

Such issues have underpinned local campaigns against the SC plan across the South and East of England. Local and regional authorities, along with environmental groups, have been successful at limiting the amount of development that has thus far taken place. The early evidence suggests that the SCs agenda is being actively challenged and reshaped through action at different scales. For example, in core development areas across the South East local actors have been drawing on EU planning directives on wildlife protection to challenge and contest development proposals, often with great success (see *The Independent*, 2006). One consequence of European Union membership is that the planning system has become, in many ways, increasingly complex and now draws on multiple priorities and objectives, set at different scales, thus opening up new opportunities for different stakeholders to oppose development proposals. This is a particular problem for the SCs plans, as they directly threaten the dwindling habitats of many species in the South East, an area that is already one of the most densely populated in Europe.

In a similar vein, the plans have also been challenged by heritage organizations. In 2004 the government agency English Heritage published a report titled *A Welcome Home—a Sense of Place for a New Thames Gateway* in which it argued that the government had adopted too much of a "blank-slate" approach to the areas identified by the SC plan and that whole areas of the South East of England were being treated simply as development sites for the building of SCs. The historical richness and diversity of such places have been undervalued, or even ignored, in the pursuit of wider developmental gains. If SCs are to be truly inclusive and improve the quality of peoples' lives, English Heritage argues, then diverse understandings and ways of thinking about specific sites need to be recognized and incorporated into development planning. There is currently little recognition that different types of communities may have very different attachments to places and specific sites as well as different needs and priorities. Building over places of cultural or even scientific interest can undermine the quality of SC environments and, paradoxically, undermine their attractiveness to potential residents and businesses.

The fact that the physical environment has played such a key role in these debates is a reflection of the limited attention given to it in the SC plan and other development blueprints. For a plan that makes so much play of its "sustainability" credentials, it is remarkable how little the environment actually features in the proposals. The failure to do so is now helping to fuel political opposition and may in the longer run undermine the plan's defined objectives and strategies. There is a wider relevance to this in that the modernist assumption that economic growth is not only an inevitability but also something that is desirable may become increasingly constrained by the physical limitations imposed by resource exploitation. Some of the new regionalist work on state accumulation strategies (see Jessop, 2002) also needs to address, in more explicit terms, the longer-term implications of such environmental limits and the effects they may have on the possibilities and capacities of state planning.

To summarize, this section has argued that the SC agenda represents a heuristic vehicle in and through which broader processes of spatial policy planning, governance, and state restructuring can be examined and assessed. The discourses and imaginations that underpin the new approaches reflect and reproduce wider shifts in spatial policy thinking. There is clearly a "new regionalist" emphasis to the plans, with their explicit concern over the competitiveness of the English regions and their propagation of globalist development ideas and discourses. They also embody a shift in spatial policy priorities, with a move away from the direct support of entrepreneurs to the indirect support of economic growth through the provision of consumption assets, such as housing. And yet, this section has also indicated that the implementation of such discourses has thus far been anything but coherent. As yet there has been little in the way of sustained public or private sector investment that can successfully "accommodate" growth in the faster-growing regions or tackle the real shortages of infrastructure supply that exist within them. Given this emerging context, the next section now turns to questions over the sustainability of the new agendas and identifies areas for further research.

THE SUSTAINABILITY OF THE SUSTAINABLE COMMUNITIES AGENDA

Despite its undisputed rise during the 2000s, the discourse of the SC is likely to undergo significant recharacterizations, challenges, and changes in the future. This section explores some of the possibilities and prospects of the new agendas and highlights some of the directions that debates over SCs may take. Although much of the discussion thus far has focused on the English case, the themes outlined in this section have a much broader resonance for spatial policy practices and rationalities elsewhere. It examines, in short, the sustainability of the SC agenda through an interrogation of emerging trends in the processes of governance, spatial policy planning, and environmental change.

The first point to make is that, in the short term, the broad thrust of the SCs plans are likely to remain high on national and international policy agendas, as they are embedded in the reforms to processes and systems of governance under way in many countries. In many Western countries, for example, there is a growing acceptance that traditional modes of *representative democracy*, particularly at the local and regional levels, are outdated and that one of the core priorities of spatial policy should be to encourage direct, active, and participative forms of democracy through the creation of SCs. In England this has, up to now, involved the implementation of a hybrid system in which elected local representatives still control local authorities and work alongside community groups, business leaders, and other interests. However, the Blair government is considering more radical reforms in which powers and resources will increasingly be devolved to identified local, or "sustainable," communities, with "old-fashioned" local government increasingly cut out of state programs and expenditure altogether. This type of direct community-based governance is, therefore, likely to be expanded and developed in the future, and there is little in the way of a "counterdiscourse" being developed by political groups at local, regional, national, or international scales. The idea of interventionist, "big" government is currently absent from the mainstream political agendas across the EU and North America, and the construction of SCs is likely to become a more attractive option that can facilitate and legitimate wider state reforms.

The broader principles inherent in the SC agenda reflect and reproduce new ways of thinking about communities, citizens, and places. An SC is one that is self-managing and "stands on its own two feet." Communities are expected to take on more responsibility for their own circumstances and to move away from any sense of "dependence" on the state. There are explicit connections here with the "good governance" principles outlined by global organizations such as the United Nations and the World Bank and the ways in which they have interpreted the concept of sustainability to legitimate and justify less interventionist, "smaller" systems of government (see Harvey, 1996; Meadowcroft, 2000). The new agendas, therefore, reflect and reproduce new ways of thinking about citizenship and what the balance of rights and responsibilities should be among different individuals, communities, and the state.

However, processes and trends in governance systems are always subject to contestation, challenge, and modification. It is possible, for example, that there will be something of a backlash against community-driven agendas and SC programs more broadly. As we have seen, the deficiencies in its implementation in countries such as the United Kingdom are likely to limit the short- to medium-term effectiveness and socioeconomic impacts of policy. The lack of compulsion on private sector developers and communities to act in a coordinated and targeted way is likely to undermine the efficacy and ultimately the legitimacy of such programs. There is a broad range of evidence that documents the limitations and problems of existing programs and strategies of community-led development programs (see for example Cochrane, 2003; Taylor, Craig, & Wilkinson, 2000). Indeed, during the 2000s what could become a countertrend has emerged in the U.K. planning systems with the introduction of new measures that call for more "strategic" agendas to be developed at the local and regional levels, in large part in reaction to the perceived fragmentation and lack of coordination within community-driven programs (see Tewdwr-Jones, 2006). It is possible that, as the form and character of governance priorities and systems are negotiated in the future, the term SC will quickly become discredited and replaced by other development discourses that distance themselves from the perceived failures of the program.

One additional factor that may influence these processes is the growing impact and significance of environmental change. Without significant reforms to existing economic and social systems, environmental pressures will inevitably increase in the future. The growing impact of global warming and resource depletion on the global environment will become increasingly obvious in the coming decades, and the intensity of the debates surrounding the meaning of sustainability could increase significantly. In this context, the form and character of SCs discourses may be subject to change, and the existing (primary) focus on the social and economic development of places may be superseded by a growing concern with the environmental impacts of urban change and regeneration. In this sense we may be witnessing the beginning of a broader set of changes in the governmentalities of communities, citizens, and policymakers so that in the long run the priorities of development programmes may be very different to growth-oriented agendas of contemporary policy. Indeed, as environmental resource issues become more significant in the future, the processes through which SC and sustainability more broadly are defined and with what purpose may become increasingly politicized.

Overall, then, the future of spatial policy and SC-building programs is closely bound up with the changing nature of the state and dominant ways of thinking about the state's role, what its boundaries and parameters of action ought to be, and what the rights and responsibilities of different citizens are and should be. If the SC program were to be fully implemented, the state's role would become that of a facilitator, intervening indirectly to support those individuals, communities, places, and regions that are best able to help themselves while encouraging others to be less "dependent" and more proactive in creating better futures for themselves. The extent to which this will be possible, however, will depend on broader social, economic, and environmental factors, and the ways in which they are (and can be) regulated and controlled are the themes to which our conclusions now turn.

CONCLUSIONS

This chapter has examined and assessed the rise of sustainable communities planning in England. It has argued that its emergence reflects and reproduces wider trends in development thinking, with its emphasis on sustainability (broadly defined) and the bringing together of social, economic, and environmental objectives. It has expanded on earlier work by highlighting some of the core elements of the new agendas and explaining how and why they became central to the Blair government's wider modernization programs. In many ways these changes reflect deeper shifts in thinking, in which it is increasingly "accepted" that the state should support its globally competitive, economically prosperous regions for the wider "benefit" of all of its regions.

There are three principal conclusions that can be drawn here.

1. The chapter has raised significant methodological and conceptual issues that are relevant to the study of spatial policy in and beyond the English context. It has suggested that too often there is a disjuncture between the analysis of discourses and representations on the one hand and the processes, practices, and politically constructed nature of policy implementation and delivery on the other. Research needs to weave these different stands together to examine the ways in which each influences and shapes the other. It is too easy to "read-off" development logics from policy documents and stated aspirations without giving due weight to the contested politics of spatial policy delivery. The chapter has shown that within the English SC agenda it is possible to identify a new

regionalist emphasis on supporting and sustaining growth in the most competitive regions. However, it would be a mistake to leave any analysis at that point. Three years into the SC program, it is difficult to see exactly what impact the proposals have had on the English regions or on the balance of socioeconomic development among them. While there has been little in the way of a return to nationally oriented Keynesian agendas, there has not been a full implementation of a selective regionalist agenda either.

2. The emergence of SCs agendas is recognizable as a trend in development policy thinking in both developed and developing countries. This reflects the spread of "sustainability" discourses into a growing range of policy areas and its "recontextualization" (see Maloutas, 2003) in different circumstances. The discourse's emphasis on democratization, new forms of equitable growth, and longer-term thinking about environmental resources has had some impact on development planning and thinking across the EU and beyond. However, the SC case has demonstrated some of the tensions involved in changing policy processes and systems to make them more sustainable. For example, despite the U.K. government's emphasis on reducing the size of the state and supporting bottom-up community initiatives (in line with some readings of the sustainability discourse), there is a growing realization that in order to deploy holistic agendas, as outlined in the government's development blueprints, new more powerful and better resourced state institutions will be required. This raises an empirical question. Can weak "hollowed-out" nation-states implement effective new regionalist agendas if those agendas require strong forms of state intervention and significant investment? In this sense the diffusion of sustainability agendas does open up new possibilities for the emergence of a new politics of state intervention, but clear tensions remain. There are also particular challenges for states across the EU and North America in which public welfare institutions are currently being restructured and the capacities of states are being undermined.

3. The chapter has also indicated that there are significant avenues for further research on spatial policy throughout the developed world. One theme for further exploration will be, for example, to look at the ways in which SCs, in different places, are defined in the future and what types of place imaginations they will reflect and reproduce. Any definition of a SC involves decisions to be made over what its boundaries should be, at what scale they should be drawn, and whose *presence* and *absence* is desirable for its functioning. But the definitions over who

is and who is not an essential person within an SC are likely to change over time as different needs and priorities are identified. In a context of enhanced globalization and structural changes to many urban and regional economies it is likely that the contours of spatial planning will change significantly in the future. As a political construction, created to play a role in broader reformist state agendas, SCs will be subject to redefinition and rearticulation in the future, even if the growing emphasis on the new regionalism and new ways of thinking about spatial policy are reproduced in the future. The same is true for broader questions over sustainability, and research will be required to examine how the sustainability discourse will be reimagined and with what effects; the types of sustainability (such as economic, social, and environmental) that will be prioritized in policy agendas and by whom; and the relationships between understandings of sustainability and spatial policy in the future and how these will reflect and reproduce broader state rationalities and the structures of regulation. Indeed, exploring the sustainability of sustainability agendas may come to represent one of the most important and fruitful areas of spatial policy and planning research.

NOTES

1. In the words of the European Commission (2006): "Sustainable development has been a fundamental objective of the European Union since 1997. It was enshrined as article 2 of the Treaty [of Amsterdam]. It is supposed to underpin all EU policies and actions as an over-arching principle."
2. A report published by the U.S. Central Intelligence Agency (2005) uses Gini Index scores to rank socioeconomic inequalities in different countries. A score of 100 equals perfect inequality, a score of 1 perfect equality. The figures for the EU in 2004 were: Austria 31; Belgium 28.7; Czech Republic 25.4; Denmark 24.7; Estonia 37; Finland 25.6; France 32.7; Germany 30; Greece 35.4; Hungary 24.4; Ireland 35.9; Italy 27.3; Latvia 32; Lithuania 34; Netherlands 32.6; Poland 31.6; Portugal 35.6; Slovakia 26.3; Slovenia 28.4; Spain 32.5; Sweden 25; *United Kingdom 36.8.*

REFERENCES

Amin, A., & Thrift, N. (2002). *Cities: Re-imagining the urban*. Cambridge, UK: Polity Press.
Barker, K. (2004). *Delivering stability: Securing our housing needs*. London: Her Majesty's Stationery Office.
Barker, K. (2006). *Barker review of planning—interim report*. Department for Commu-

nities and Local Government, London. Accessed at *http://www.communities. gov.uk/index.asp?id=1501326*

Brenner, N. (2003). Globalisation as a state spatial strategy: Urban entrepreneurialism and the new politics of uneven development in western Europe. In J. Peck & H. Wai-chung Yeung (Eds.), *Remaking the global economy—economic–geographical perspectives* (pp. 197–215). London: Sage.

Brown, G. (2006). *Budget statement 2006* (speech to House of Commons). Accessed March 22, 2003, at *http://www.hm-treasury.gov.uk/budget/budget_06/ bud_bud06_index.cfm.*

British Broadcasting Corporation. (2006). New homes fear over water crisis. Accessed April 20, 2006, at *http://www.news.bbc.co.uk/1/hi/england/4928352.stm.*

Burton, E. (2000). The compact city: Just or just compact? A preliminary analysis. *Urban Studies, 37,* 1969–2006.

Campaign for the Protection of Rural England (CPRE). (2004) *The Thames Gateway— making progress.* London: Author.

Central Intelligence Agency. (2005). *The world fact book 2005.* Available at *http:// www.cia.gov/cia/publications/factbook/fields/2172.html.*

Cochrane, A. (2003). The new urban policy: Towards empowerment or incorporation?. In R. Imrie & M. Raco (Eds.), *Urban renaissance?: New Labour, community and urban policy* (pp. 223–234). Bristol, UK: Policy Press.

Counsell, D. (1999). Sustainable development and structure plans in England and Wales: Operationalising the themes and principles. *Journal of Environmental Planning and Management, 42,* 45–61.

De Roo, G., & Miller, D. (2000). *Compact cities and sustainable urban development— a critical assessment of policies and plans from an international perspective.* London: Ashgate.

The Economist (1945, March 3). The distribution of industry, pp. 270–272.

English Heritage. (2004). *A welcome home—A sense of place for a new Thames Gateway.* London: Author.

English Nature. (2006). *Sustainable development report, 2004/2005.* London: Author.

Environment Agency. (2005). *Creating a better place—the Environment Agency's assessment of the environment in South East England.* London: Author.

European Commission. (2006). *Sustainable development: EU strategy.* Available at *www.euractiv.com/en/environment/sustainable-development-eu-strategy/article-117544.*

Gamble, A. (1988). *The free economy and the strong state.* Cambridge, UK: Polity Press.

Gibbs, D. (2002). *Local economic development and the environment.* London: Routledge.

Giddens, A. (2005). Neoprogressivism: A new agenda for social democracy. In A. Giddens (Ed.), *The progressive manifesto: New ideas for the centre left* (pp. 1–35). Cambridge, UK: Polity Press.

Gordon, I., & Buck, N. (2005). Introduction: Cities in the new conventional wisdom. In N. Buck, I. Gordon, A. Harding, & I. Turok (Eds.), *Changing cities—Rethinking urban competitiveness, cohesion and governance* (pp. 1–24). London: Palgrave.

Harvey, D. (1996). *Justice, nature and the politics of difference.* Oxford, UK: Blackwell.

Harvey, D. (2005). *A history of neoliberalism.* Oxford, UK: Blackwell.

Heseltine, M. (1990). *Where there's a will.* London: Hutchinson Press.

Houlihan, B. (1988). *Housing policy and central–local government relations.* Hampshire, UK: Avebury Press.

Imrie, R., & Raco, M. (2003). *Urban renaissance?: New Labour, community and urban policy*. Bristol, UK: Policy Press.

The Independent. (2006, May 2). The birds that blocked 20,000 homes, p. 1.

Jenks, M., Burton, E., & Williams, K. (Eds.). (1996). *The compact city: A sustainable urban form?* London: E & F Spon.

Jessop, B. (2002). *The future of the capitalist state*. Cambridge, UK: Polity Press.

Jones, M. (1997). Spatial selectivity of the state. The regulationist enigma and local struggles over economic governance, *Environment and Planning A, 29*, 831–864.

Lawless, P. (1989). *Britain's inner cities*. London: Paul Chapman Publishing.

Livingstone, K. (2004). "Foreword." In Greater London Assembly, *The case for London—London's loss is no-one's gain* (pp. 3–4). London: Author.

Lovering, J. (2006, March 2–3). Regionalism and the neo-liberalisation of governance. Paper presented at the Economic and Social Research Council seminar *The rise of multi-level governance and meta-governance in an international context*, The University of Manchester.

Maloutas, T. (2003). Promoting social sustainability: The case of Athens. *City, 7*, 165–179.

Meadowcroft, J. (1999). Planning for sustainable development: What can be learnt from the critics? In M. Kenny & J. Meadowcroft (Eds.), *Planning for sustainability* (pp. 12–38). London: Routledge.

Meadowcroft, J. (2000). Sustainable development: A new(ish) idea for a new century? *Political Studies, 48*, 370–387.

Office of the Deputy Prime Minister. (2003). *Sustainable communities: Building for the future*. London: Her Majesty's Stationery Office.

Office of the Deputy Prime Minister. (2004). *Planning policy statement 1: Planning for sustainable development*. London: Her Majesty's Stationery Office.

Office of the Deputy Prime Minister. (2005). *Sustainable communities: Building for the future*. Available at http://www.dclg.gov.uk.

Pearce, D., Markandya, A., & Barbier, E. (1991). *Blueprint for a green economy*. London: Earthscan.

Peck, J., & Tickell, A. (2002). Neo-liberalising space. In N. Brenner & N. Theodore (Eds.), *Spaces of Neoliberalism: Urban Restructuring in North America and Western Europe* (pp. 33–57). Oxford, UK: Blackwell.

Raco, M. (2005). Sustainable development, rolled-out neo-liberalism and sustainable communities. *Antipode, 37*, 324–346.

Raco, M. (2007). *Building sustainable communities—Spatial policy, place imaginations, and labour mobility in post-war Britain*. Bristol, UK: Policy Press.

Rose, N. (1999). *The powers of freedom*. Cambridge, UK: Cambridge University Press.

Schuppert, F. (2005). The ensuring state. In A. Giddens (Ed.), *The progressive manifesto: New ideas for the centre left* (pp. 54–72). Cambridge, UK: Polity Press.

Scott, A. (2005). *Global city-regions: Trends, theory, policy*. Oxford, UK: Oxford University Press.

South East England Development Agency. (2003). *Board meeting Wednesday 11 June—Item 8 Housing*. Guildford, UK: Author.

Taylor, M., Craig, G., & Wilkinson, M. (2000). Co-option or empowerment?: The changing relationship between the state and the voluntary and community sectors. *Local Governance, 28*, 1–11.

Tewdwr-Jones, M. (2006). Identifying the determinants of the form of government, governance and spatial plan making. In M. Tewdwr-Jones & P. Allmendinger

(Eds.), *Territory, identity, and spatial planning* (pp. 353–365). London: Routledge.

Ward, S. (2004). *Planning and urban change* (2nd ed.). London: Sage.

While, A., Jonas, A., & Gibbs, D. (2004). The environment and the entrepreneurial city: Searching for the urban sustainability fix in Manchester and Leeds. *International Journal of Urban and Regional Research, 28,* 549–569.

Whitehead, M. (2003). (Re)analysing the sustainable city: Nature, urbanisation, and the regulation of socio-environmental relations in the U.K. *Urban Studies, 40,* 1183–1206.

CHAPTER 9

The Spatial Politics
of Conservation Planning

JAMES P. EVANS

Over 100 nations signed the Convention on Biodiversity at the United Nations Conference on Environment and Development in 1992. It was a potentially historic moment. Biological conservation had been given a global footing, levered into national, regional, and local agendas on an unprecedented magnitude through the concept of sustainable development. The inclusive definition of biodiversity[1] explicitly extended nature conservation to cover all areas of human habitation, adding urban and suburban areas to the rural and wilderness habitats more familiar to conservation activities.

This move was timely. Many of the world's biodiversity hotspots have higher-than-average population densities, and rapid rates of urban and suburban sprawl are threatening biodiversity in the Americas, Australia, and southern Africa (Mittermeier, Myers, & Mittermeier, 2000; United Nations Center for Human Settlements, 1996). Even areas characterized by low levels of human population density are not beyond the reach of human impacts. As Bruce McKibben (1999) argues in *The End of Nature*, no part of the planet remains untouched by pollution, from the most distant Arctic tundra to the upper atmosphere. At the same time as an ever increasing proportion of the world's human population is living in suburban or urban environments, so these environments are being colonized by a variety of species (Mabey, 1999). To take two examples from many possible, 1.5 million Mexican freetail bats roost under the Congress Avenue bridge every summer in Austin, Texas, and bridges and culverts across the United States provide extensive habitat for 25 species of bats (Keeley & Tuttle, 1996). In the United Kingdom, the sheer faces of tall office buildings and rubble piles on derelict sites in city centers provide analogues of the Alpine scree slope habitats of the endangered black redstart and peregrine falcon (Evans, 2003; Hinchliffe & Whatmore, 2006). These trends present an increasing need to conserve nature in the midst of human activity (McDonnell & Pickett, 1993). In response to this challenge, biodiversity conservation is couched within the broader category of sustainable development,[2] which aims to integrate environmental, economic, and social priorities in order to achieve balanced and equitable development.

Planning is critical to the implementation of sustainable development, and a variety of movements have emerged that are concerned with how best to organize development, from "smart growth" in the United States to the livable cities agenda in the United Kingdom. The current era of environmental planning is characterized by an extension and intensification of our efforts to regulate and administrate the environment. From the beginnings of life itself until the industrial era, the environment was the medium through which we lived and acted. In the postindustrial era the environment itself has become an object, to be manipulated through the medium of technical management and planning. Along with "genetic monitoring, wildlife tourism, GIS [geographic information system] tracking, and myriad other technologies, conservation planning administrates an increasingly dense set of relationships" between humans and the environment (Whatmore, 2002: 13). Decisions must be made and enforced concerning the relative worth of animals, habitats and supermarket jobs, between nature reserves and retail parks. Put simply, space is under pressure to accommodate more people, animals, and things than ever before.

In the face of these regulatory dilemmas, politicians and planners often turn to conservation science to add legitimacy to decisions. Under pressure to plan more sustainably, interest in the fields of landscape ecology and conservation biology has blossomed.[3] Conservation models are of particular use, as they explicitly address how best to plan human and ecological features across an area, helping planners to decide, for example, whether it is more effective to completely section conservation areas off from areas of human development, what properties these areas should have in terms of size, number, proximity, and connectivity, and so forth. Conservation models emphasize the need to retain habitat across landscapes fragmented by urban and rural land uses through strategic planning (Collinge, 2001).

Conservation planning is thus an important arena in which sustainability plays out. The way in which we resolve questions concerning who and what belongs where, how and where they should be allowed to mix, and, in the final analysis, what kinds of future landscapes that we want to create depends upon how we view our relationship with nature. As planning tools conservation models are intriguingly placed on the cusp of ecology and planning, closely related to the life sciences but also implicated in wider political and economic processes of development (Gandy, 2002; Swyngedouw, 1999). As part of the wider institutional framework of planning they can be "hegemonic or anti-hegemonic, supportive of existing regulatory structures or counter-regulatory" (Desfor & Keil, 2000: 5). In turning cultural relations into spatial distributions of nature and humans, conservation planning is a politically charged activity, generating a series of social consequences (Harvey, 1996).

This chapter explores what I term the "spatial politics" of conservation planning. I use this term to emphasize the simultaneously geographical and political character of conservation planning and to interrogate the relations among space, politics, and science. These debates are explored through the example of the dominant "fragmentation" model of conservation planning. While questionable on scientific and political grounds, the fragmentation model persists in policy and practice. In the face of calls to radically shift the scales at which landscapes are planned, I explore the causes of this persistence, from the genealogy of conservation biology out of which it developed through to actual case studies of how the model is deployed in practice.

The first case examines local political resistance to the European Natura 2000 reserve network in the rural Finnish context, while the second traces how the fragmentation model frames the trade-off between urban conservation and development in the United Kingdom. Through these examples I explore the relationship between the spatial politics of conservation planning and the exercise of administrative power. I conclude by discussing the wider implications that this critique of conservation planning may hold for the spatial politics of sustainability.

THE FRAGMENTATION MODEL
AND ITS ECOLOGICAL CRITIQUE

A number of branches of spatial ecology inform conservation planning. Landscape ecology, conservation biology, and biogeography are all concerned with the spatial application of biological principles to generate models of conservation planning, from specific reserve design to landscape planning. Conservation biology involves the protection and management of biodiversity, referring to both the science and application of that science to address a variety of ecological problems, ranging from specific populations of endangered species to regional conservation planning. Landscape ecology tends to be concerned with the ecological conceptualization of entire landscapes, focusing on the causes and consequences of spatial heterogeneity (Forman, 1995), and biogeography is closely related, studying the geographical expression of biological processes. The arguments of this chapter focus on how ecological conceptions of landscape are used to inform conservation planning, and thus draw most heavily on landscape ecology, although as will become apparent these subdisciplines overlap considerably.

The three traditional components of landscape ecology are the patch, the edge, and the corridor (Duane, 2000). As Figure 9.1 shows, corridors are linear features that differ from the surrounding landscape, linking habitat areas that were once historically connected (Peck, 1998). The boundaries of corridors and patches are called edge habitat. Landscape ecology supposes that a network of linear green corridors linking otherwise isolated areas of habitat across a fragmented landscape will maintain higher levels of biodiversity (Barker, 1997; Forman, 1991) than one in which no links exist. The connectivity of habitat they offer species in terms of movement and resources is seen as a necessary ecological corrective to the excessive fragmentation of habitat that characterizes anthropogenically modified landscapes (Collinge, 1996; Dunning, Danielson, & Pulliam, 1992).

The idea of habitat fragmentation originated in MacArthur and Wilson's theory of island biogeography (1967), which sought to explain species diversity on neighboring islands in the sea. In this work, they suggest that the principles of "insularity" (islandness) apply universally to landscape elements like forests, streams, and so on. They claim, furthermore, that insularity will apply increasingly to habitats being broken up by human land uses, as nature reserves and remnant habitat patches become like islands within the hostile surrounding landscape. Habitat

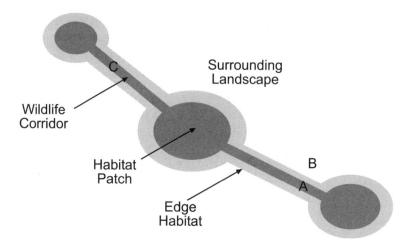

FIGURE 9.1. Elements of conservation planning.

fragmentation has become a central theme for conservation biologists seeking to understand and describe the "human-induced deterioration of the environment" (Haila, 2002: 321). As the Finnish ecologist Yrjö Haila notes, what is essentially a single paragraph from their monograph became the basis for the discipline of landscape ecology.[4]

The notion of linked sites forms the basis for the European network of biogenetic reserves—the Natura 2000 network, which seeks to maintain key habitats in urban, rural, and wilderness contexts under the Habitats Directive (European Union, 1979) in line with the Convention on Biodiversity (Glowka, Burhenne-Guilmin, & Synge, 1994). Wildlife corridors and habitat patches underpin the North American landscape conservation programs, from the design of individual national wildlife reserves to the Wildlands Project that aims to reconnect the major biomes of North America. Corridors are also used in individual reserve design. For example, the Lower Rio Grande Valley National Wildlife Refuge in southern Texas, which was established to preserve the valley's native vegetation in the face of increasing residential and industrial development, is organized around a 275-mile river corridor (Peck, 1998). The importance of linking sites has been highlighted at the World Summit on Sustainable Development held in Johannesburg in 2002, the World Parks Conference in Durban in 2003, and the Convention on Biodiversity Conference of Parties 7 in 2004 (FERN, 2004).

In his review of fragmentation thinking within landscape ecology,

Haila (2002) argues that the idea came to prominence as part of the more general environmental awakening in the 1960s and 1970s, when there was widespread demand for new perspectives on ecological problems. As he says, "Fragmentation research was thought to provide insight on human-induced extinctions through the analogy between species impoverishment on islands and habitat fragments" (Haila, 2002: 327). The concept of fragmentation was timely in that its initial emergence coincided with an upsurge in popular concern over ecological degradation that accompanied the rise of environmentalism. A subsequent wave of professional interest can be identified in the post-Rio context; as environmentalism became assimilated into the political mainstream, planners were suddenly expected to incorporate strategic conservation into their activities. Decision makers like to believe that ecology is "socially . . . [and] politically innocent" (Hiedanpää, 2002: 122), an objective scientific domain untainted by pragmatic constraints that can provide incontestable "answers" to planning dilemmas. Landscape ecology held this promise for planners. The increasing influence of conservation planning can be seen as part of a wider proliferation of claims to ecological authority that has occurred under the science-driven rubric of sustainable development (Eden, 1999; Murdoch & Lowe, 2003). In the face of these demands, the fragmentation model became formally established as a basis for conservation planning in urban, exurban, and wilderness contexts (Fabos & Ryan, 2004).

Despite its dominance in planning policy, the scientific basis of the fragmentation model has received increasing criticism from ecologists. Three assumptions have attracted particular scrutiny. First the model confuses the process of habitat fragmentation with the process of habitat loss. Much of the original research upon fragmentation and species change argued that, while habitat loss to, for example, agricultural expansion might impact negatively upon biota, the edge habitat created may bring new species into a region. When the model of island biogeography was applied to these studies, it was assumed that habitat fragmentation would inevitably lead to a loss of species. Second, the model assumes a reference landscape prior to fragmentation that is characterized by a uniform and continuous habitat. As Haila notes, this "spatially homogenous and temporally constant natural background" (2002: 326) represents an ideal that neglects the temporal and spatial variations in biotic elements across a landscape. Third, the fragmentation model assumes that areas surrounding fragments of habitat are entirely inhospitable as ecological environments. This claim can be interrogated empirically but is "untenable as a broad generalisation" (Haila, 2002: 325). Not only is the environment surrounding habitat fragments heterogeneous (different types of agriculture, suburbia, etc.), but its relative amenability to different species will also vary massively.

The long-running debate surrounding the utility of wildlife corridors linking habitat patches is indicative of these problems (Barker, 1997; Beier & Noss, 1998; Dawson, 1994). On the one side lies a wealth of anecdotal and intuitive evidence supporting their efficacy (Spellerberg & Gaywood, 1993). However, little hard scientific evidence exists to prove that linear connectivity creates higher levels of biodiversity in a fragmented landscape (Dawson, 1994; Nicholls & Margules, 1991). Research upon the ecological efficacy of wildlife corridors has been hindered by the difficulties of testing dynamic dispersal attributes for both individual species and communities of organisms (Adams & Dove, 1989). Species-specific research has shown the importance of various surface types to movement, but often what is a conduit for one species will be a barrier for another (Yanes, Velasco, & Suárez, 1995). The lack of ecological evidence has led one commentator to claim that corridors "appeal more as intuitive constructs than they do as a set of scientifically tested findings" (Boothby, 2000: 283). As Suzy Becker's cartoon in Figure 9.2 shows, planners can impose corridors, but whether species actually use them is another matter.

The three criticisms of the fragmentation model all question the underlying assumption that human-influenced environments are essentially and always "hostile" to natural elements and that preexisting "referent" or "natural" ones are stable ecological oases. Human-influenced environments are ecologically less stable and thus assumed to be hostile, while referent landscapes are assumed to be harmoniously stable and thus uniformly amenable to ecological elements (Haila, 1999b). The idea that human-influenced environments are essentially different from so-called natural ones is at best a gross oversimplification and at worst empirically unfounded (Whitmore, 1997). There is evidence that the matrix surrounding habitat fragments is actually porous to many biotic elements, undermining the core assumption of insularity (Small, Sadler, & Telfer, 2003; Whitfield, 2001). The notion of habitat fragmentation is not only empirically questionable but also politically loaded. In assuming that nature is generally absent from landscapes that are highly modified by humans, the fragmentation model presupposes that effective conservation should spatially segregate humans and nature (Haila, 1999a,

FIGURE 9.2. The wildlife corridor as a cultural construct. From Becker (2002). Copyright 2002 by Suzy Becker. Reprinted by permission.

2000). The spatial politics of this underlying assumption fly in the face of the integrative goals of sustainable development.

The causes of this disjuncture between the demands of sustainable planning and traditional landscape ecology are partly historical. While the motives of conservation planning can be related back to the earliest efforts of humans to manage and shape their environment, the distinct set of regulatory practices that are familiar to us today emerged from efforts in the 18th and 19th centuries to plan and protect natural areas such as national parks against urbanization and industrialization (Adams, 1986). Ecology developed as a field science that used these relatively untouched areas as outdoor research laboratories. Not only did ecologists work in landscapes relatively free from human influence, but they tended to focus on small areas such as sand dunes and lakes, where it was easier to define clear boundaries for distinct ecosystems (Worster, 1977). These studies generated powerful concepts, among which the model of ecological succession is probably the most well known. This concept suggests that under stable conditions ecosystems will increase in complexity toward a climax state, and conversely that the effects of increasing human activity such as building, agriculture, or transport links upon a landscape will limit succession and biodiversity. Because fragmentation disrupts the equilibrium of a landscape, it is assumed to cause habitat loss. These models reflect the landscapes in which they were devised, tested, and refined—landscapes characterized by a lack of human influence. The traditional models of ecology break down when they are applied to human modified landscapes that are characterized by disturbance and disequilibrium as it becomes impractical and undesirable to separate ecological and anthropogenic processes.

This bias persists. While the holistic aspirations of sustainable development and biodiversity conservation seek to encourage and conserve biodiversity where people actually live, conservation biology tends to neglect areas characterized by higher levels of human settlement. James Miller and Richard Hobbs (2002: 331) reviewed papers published in *Conservation Biology* from 1995 to 1999 in order to gauge the amount of research being conducted in urban, suburban, and exurban environments. They found that less than 6% of the papers considered human settlement in any way, supporting their contention that conservation biologists have tended to neglect the issue of conservation in landscapes highly modified by human settlement. What we have, essentially, is a disjuncture between the current policy demands of sustainability placed upon conservation planning in urban and suburban environments and the tools that it possesses to address this challenge.

BEYOND FRAGMENTATION?: INTEGRATION, SCALE, AND PERSISTENCE

The spatial politics of the fragmentation model raises two immediate questions: first, whether the model is suited to the integrative goals of sustainable planning and, second, whether it is the most ecologically legitimate mode of conservation planning. In response to this disconnect, commentators have suggested that the sustainable management of ecological resources may require us to shift scales in order to integrate heterogeneous environments (Lee, 1993). Consequently, we see advocates from either end of the landscape ecology spectrum trying to colonize new scales. Ecologists have attempted to produce ecological models that integrate processes at political scales, such as the city, while political analysts have criticized the continuing use of political scales that frustrate ecological units such as watersheds or biomes. The notion of scale underpins the rhetoric of integration informing current calls for sustainability, and it is worth exploring examples of each.

Conservation biologists have developed models that accommodate the heterogeneity of anthropogenically altered landscapes. Landscape ecology has largely left the island biogeography version of the fragmentation model behind for a "new paradigm" of metapopulation studies (Pickett, Parker, & Fiedler, 1992), focusing upon "the determinants of population viability in heterogeneous and variable non-equilibrium environments" (Haila, 2002: 329). For example, urban ecologists have expended considerable effort attempting to reconceive cities as fully functioning socioecological entities rather than as ecological vacuums. Pickett, Cadenasso, and Grove (2003: 373) argue that urban areas can be thought of in terms of ecological resilience, or their ability "to adapt and adjust to changing internal or external processes" (see also Holling, 1973). Drawing heavily on nonequilibrium ecology, humans are conceptualized as an integral part of the "resilient city ecosystem"—a system that is open to energy and materials, characterized by disturbance, and that may have multiple stable states. The idea of resilient cities is explicitly concerned with reframing the spatial assumptions of classical ecology away from the discrete homogeneous spaces of equilibrium (those lakes and sand dunes), toward the acknowledgment of spatial heterogeneity as a governing force at a range of scales (Pickett & Cadenasso, 1995).

Understanding cities in this way can allow planners to link ecological and social heterogeneity at the strategic level, deploying different ecological models to plan rapid and spatially uneven development. Using the examples of watersheds, they argue that ecological, social, and infrastructural functions can be integrated and planned in a spatially heterogeneous way. Pickett's discussion hinges on the idea that ecology and planning can be joined together in terms of how they articulate space, specifically through the manipulation of diversity within certain scales in order to incorporate different forms of land use. In order to do so, some commentators have begun championing alternative spatial ontologies that release nature conservation from the "ghettos" (Boothby, 2004: 67) of protected sites. Boothby argues that "the definition of boundaries is central to the meanings and concepts within a planning system, for they demarcate where one space ends and where another begins" (2004: 74). Drawing on Smith (1995), he suggests that the notion of "incomplete boundaries" can be of use to conservation planning in outlining a system that can evolve holistically within its local context. Rather than the either/or linear separation of the fragmentation model, this system would overlay nature conservation on to other land-use priorities.

A similar preoccupation is found in political ecology. Karl Zimmerer

(1994, 2000) has suggested that conservation geographies may be re-worked through nonequilibrium ecology to focus on the creation of less segregated landscapes that seek to recognize rather than efface the close relations between society and nature. Bruce Braun (2005) has recently asked what the spatial category "urban" adds to any consideration of urban environmental issues, arguing that the urban has little meaning as an object of ecological analysis and the ecological unbounding of the city is a crucial step toward integrating planning and ecology. Quoting Haughton (1999: 233, in Braun, 2005: 639), he argues that the process of disintegrating the city as a conceptual container is essential to estab-lishing a meaningful basis for the concept of sustainable cities. "[A] sus-tainable city cannot be achieved purely in internal terms. . . . It is futile and indeed virtually meaningless to attempt to create a 'sustainable city' in isolation." Echoing the arguments of Pickett and Cadenasso (1995), his analysis of sustainability also settles on a spatial prognosis—but rather than articulate the urban as an ecological level, he suggests that we reject the city as a spatial unit in order to integrate scales of political and ecological planning.

Both scientific and political critiques recognize the pressing need to integrate political and ecological scales of planning in order to achieve more sustainable landscapes. But while landscape ecology has moved on from equilibrium models, conservation planning has not.[5] The problems with the underlying ecology of fragmentation have not brought about its demise in planning policy. Given the flimsiness of its ecological basis, we are prompted to ask why it persists so strongly. Reflecting upon this par-adox, Haila suggests that conservationists feel the urge to use the frag-mentation model as a predictive tool in the absence of direct measure-ments, a suggestion echoed by Wilson himself (1992: 263). But using the model as an approximation remains problematic, as it smuggles a set of unwarranted and pejorative assumptions concerning the relation be-tween human influence and ecological worth into the debate. The ques-tion of why the fragmentation model persists is down to more than sci-entific instrumentalism, and involves the actual pragmatics of planning. Let's return now to the discussion of scale in order to temper some of the positions set out earlier.

Both scientific and social scientific authors emphasize the dynamic emergent properties of scale (Brenner, 2001: 592) or, to put it another way, how ecological and political processes interact to produce specific scales of conservation that are "never set, but are perpetually disputed, re-defined, reconstituted and restructured in terms of their extent, content, relative importance and interrelations" (Heynen & Swyngedouw, 2003: 912–913).[6] But are the ecological bases of governance really that malleable? Evidence abounds suggesting that the manipulation of scale in practice can be problematic. The modest achievements thus far of regional planning as a new tier of governance below the national level is one area in which a new scale of governance was expected to articulate more sustainable planning. Thomas Feldman and Andy Jonas (2000) discuss the Southern California Natural Communities Conservation Planning framework to implement a regional conservation system to conserve endangered species. They suggest that a key issue involves the difficulty of incorporating local property rights within regional frameworks for conserving biodiversity (Hurley, Ginger, & Capen, 2002, make similar arguments). Attempts to manage resources at the river basin level have been confounded by preexisting socioeconomic and political scales of governance (Sneddon, 2002). For example, the problems of implementing the European Water Framework Directive, which is intended to set up strategic watershed planning, indicate that establishing new scales of regulation is not an easy task (White & Howe, 2003).

What if we can't, as Braun suggests, simply "abandon" the urban as a unit of analysis, because the urban is already articulated as an ecological unit? What if the "urban" as a spatial unit is complicit with fragmented space? What if the current scales of regulation, far from being disconnected, are highly synchronized through a range of practices? Sustainability is not inscribed onto a blank canvas, but involves the reworking of preexisting knowledge and practices, and in relation to conservation planning there is a need to describe and understand how moments of scale are already established and negotiated in practice. There is a need to explore whether certain models are more amenable to institutional power than other more ecologically robust tools, and what the political consequences of this are. Let's now look at how the spatial politics of fragmented space plays out in practice.

FRAGMENTED SPACE IN PRACTICE

Natura 2000: Continental Conservation

Natura 2000 is the flagship conservation initiative of the European Union that aims to create a coherent network of habitats across the continent. Based on the European Union Bird Protection Directive (European Union, 1979) and Habitat Directive (European Union, 1992), Natura

2000 extends the principles of the fragmentation model of landscape planning to the continental scale. Although the establishment of corridors is not legally binding under the Habitats Directive, Natura 2000 emphasizes the importance of connectivity between Special Areas of Conservation (SACs). It is already estimated to cover 13% of the European Union's land area, although completion is not scheduled until 2010 (FERN, 2004). Within the framework of the European Union, each member state is (individually) responsible for implementing its own part of the network. This process is unavoidably political, and Natura 2000 has generated a series of contradictions between national conservation priorities and local socioeconomic realities. Juhas Hiedanpää (2002) has described the process of grassroots resistance that developed in southwest Finland in response to Natura 2000 plans for reserves and controlled areas. Opposition was particularly strong in the rural municipality of Karvia, where in 1997 four landowners staged a hunger strike, prompting a ministerial visit and the subsequent withdrawal of half the plans for the area. It is worth examining this case in more detail.

Almost 70% of Karvia consists of bogs and marshes, and much of the population depends upon the remaining areas that have been drained to allow commercial forestry. Forests are Finland's most valuable resource, and there is a long history of government involvement with the forestry industry, from encouraging greater production from the early 20th century onward, to the subsequent shift toward more ecologically oriented management (Jokinen, 2006). In Karvia, this shift has been expressed in the tensions between the primary interests of the landowners toward expanding productive land through drainage and the increasing conservation restrictions that have been extended to the bogs and marsh areas.

The Natura 2000 proposals were poised to reconfigure the spatial formation of human and ecological elements, organizing and restricting human influences through a system of reserves and corridors. Although Natura 2000 guidance recognizes the need to draw up plans to manage potential conflicts between economic and environmental priorities, it states that nature conservation values should be prioritized over economic activities in reserve areas (FERN, 2004). Tighter controls over local forestry practices were seen to be an imposition upon local practices and freedoms by external administrators and environmental authorities, constraining felling and economic opportunity. As Hiedanpää (2002: 121) states, "Given the complexity of the web of social and economic interactions that constitutes forest economy in SW [southwest] Finland, a sustainable approach to forestry and environmental management calls for a planning process which emulates the structure and dynamics of local and regional social and economic systems."

Landowners disputed the exact boundaries for the proposed reserve areas, questioning the objectivity of the ecological designations of certain areas over others. The notion that it is possible to clearly divide a landscape into land that may be farmed and land where farming operations are restricted by the use of lines drawn on a map was seen as not only an encroachment on land ownership but also a somehow arbitrary imposition. The process of bounding a landscape with lines on a map is "akin to ownership; real or perceived issues of ownership generate (often unspoken) tension across the arena of rural affairs" (Boothby, 2004: 73). The associated reordering of human activities in space through regulatory mechanisms has concrete effects upon "the material conditions, that is, public, economic and ecological conditions of regional and local life" (Hiedanpää, 2002: 115). The regulations of Natura 2000 threatened to separate foresters from local forests in order to connect the forests into a national conservation system. The complete separation of the proposals from their local context flies in the face of the avowedly local modus operandi of sustainability. Interestingly, the highly developed technologies of monitoring and regulation that had built up over time to facilitate the government's productivist agenda for the forestry industry enabled the rapid imposition of a nature conservation agenda in the region.

Natura 2000 sought to bring distant ecological entities in Karvia closer to other managed spaces of nature, allowing the central conservation authorities to control the forests of Karvia. The spatial ordering of the fragmentation model resonates with the administrative model of state control. The next section explores this argument further, taking an example from urban planning in which the fragmentation model was embedded within the dominant regulatory system.

Urban Conservation Planning

Urban environments constitute classically fragmented landscapes. Extreme modifications to land cover such as roads and buildings replicate the conditions of island biogeography more accurately than most landscapes. As the urban environmental agenda has come into focus over the past 30 years, urban and suburban conservation planning has been largely based around fragmentation concepts, whether articulated as "green networks," "greenway planning," or multifunctional corridors, in order to maintain connections among remaining "islands" of habitat in the city (Barker, 1997; Little, 1990).

This is certainly the case in Birmingham, Britain's second-largest city. Having experienced serious economic decline in the late 20th century, it is now home to the most ambitious urban regeneration program undertaken in Europe since the Second World War. Wildlife corridors are the key conservation planning tool within the council's Nature Conservation Strategy (Birmingham City Council, 1997), forming the basis for strategic ecological conservation at the city level. The conservation strategy recognizes major wildlife corridors and habitat patches that form nexuses at the confluence of corridors, and the strategy forms supplementary planning guidance, constituting a "material" (legally binding) consideration in regard to planning proposals. The following case study explores how the fragmentation model framed the ongoing development of an ecologically valuable site within the city known as Vincent Drive.

Vincent Drive lies 2 miles to the south of the city center in the district of Selly Oak, 4 miles from the city's periphery. It was the largest remaining seminatural greenspace in south Birmingham, occupying approximately 30 hectares of floodplain to the north and south of the Bourn Brook River. By 1950 the site was distinguishable as a discrete island of green space, enclosed by housing, and was designated a Grade C Site of Special Scientific Interest (SSSI) by the Nature Conservancy Council for the West Midlands (now English Nature) in 1982. The legacy of old gardens, lack of management, and varying drainage has created an array of habitats on Vincent Drive. Remnant medieval deer park and pasture is over 1,000 years old (Slater, 2002), while the Bourn Brook is characterized by mature woodland and wetland habitat along its eastern stretch.

Recent development proposals involve building a £300-million hospital that will cover approximately four-fifths of the site. A major supermarket chain has agreed to finance a new link road traversing the floodplain, canal, and railway and alterations to major road junctions at either end and to enhance the Bourn Brook wildlife corridor in return for development rights over the southern half of the site (see Evans, 2004, for a detailed account). The wildlife corridor was critical in these proposals, mediating between development and conservation priorities. The Nature Conservation Strategy recognizes the site as a nexus where four corridors meet, precluding development upon Vincent Drive. A variety of planning documents and local consultation documents produced about Vincent Drive represent it as a prime (potentially wasted) development opportunity. The ecological worth of the site as a habitat in its own right is played down, while its importance as a wildlife corridor is emphasized. The wildlife corridor was deployed as a form of ecological politics, forming an attractive "storyline" for developers, planners, and conservationists across a suite of planning policies and development plans.

The wildlife corridor was established through a variety of technical practices, ranging from the production of maps and ecological surveys to public consultations. The Phase II ecological survey completed as part of the Environmental Impact Assessment represented the ecological worth of the site as a linear corridor feature, determining the resultant spatial balance struck between development and conservation. Having been established in the surveying stage, the main mitigation proposal involved the retention of a 50-meter-wide wildlife corridor running along the Bourn Brook. The planned development on Vincent Drive trades the loss of 90% of the habitat in return for the protection of the Bourn Brook corridor.

The corridor in this case allowed planners to "ring-fence the best and trade-off the rest" (Selman, 2002: 284). Local conservation NGOs, such as the Wildlife Trust, who attempted to articulate Vincent Drive's local ecological importance to protected species for feeding and breeding were out of step with the dominant spatial practices of planning. The fragmentation model articulates a highly alluring arrangement of priorities in terms of *what* constitutes nature and *where* it should be. The ability to bind nature into discrete or separate spaces (in this case a linear space) was critical in formatting the space for development.

THE SPATIAL POLITICS OF FRAGMENTATION

In order to understand the persistence of fragmentation planning I want to explore three specific questions raised by the examples above: first, how the fragmentation model was actually implemented; second, how it facilitated spatial control; and third, what the effects of this were.

As argued in the first half of the chapter, the fragmentation model spatializes a familiar wider cultural dichotomy between humans and nature, constantly producing and reproducing discursive categories through the reordering of humans and nature. This is a relational achievement in that the category of nature, or the nonhuman,[7] is always predicated upon the category "not-nature," or "human." Within this understanding, what constitutes "wildlife," "nature," or "ecology" is not pregiven but rather made through networks of people, things, and knowledge (Whatmore, 2002: 14). These practices can be conceptualized as "spatial formations," that is, ordering networks that are established through numerous *spatial* practices (Thrift, 1996: 47). Sociospatial relations between what is taken to constitute humans and nature are produced and solidified by a constellation of technical practices, from ecological surveying and monitoring to political decision making. The two examples contrast the success of implementing the fragmentation model. Natura 2000 constituted a new scale of conservation in the European context that was not embedded in local spatial practices, resulting in considerable local resistance to the plans. In the second case, the idea of corridors was deeply embedded within the formal system of city planning, supported by an array of practices that established its reality. In the face of this, local protest, less vociferous anyway, was swept aside. It is the performance of ecology through "heterogeneous social networks" (Whatmore, 2002: 14) that produces the space of the conservation model.

So, in arranging humans and nonhumans in specific ways, conservation technologies produce relational spaces of humans and nature. But how does this mode of organizing space facilitate political control? Taking the fragmentation model, the idea of insularity is predicated upon a relational logic between a homogeneous inside that is defined in opposition to a homogeneous outside. Insularity is achieved literally through "bounding" space with lines, connecting that which is supposedly the same (habitat) from that which is supposedly different (surrounding landscape). It gathers points on the inside together by bounding them off from the hostile surrounding sea. Internal connectivity is achieved through the process of linear segregation. This mode of ordering can be seen as a form of *topological* space, space that is ordered by the relations between things (in this case humans and nature) rather than through the absolute coordinates of Euclidean space (Mol & Law, 1994). The motivating insight behind topology is that some geometric problems depend not on the exact shape of the objects involved but rather on the ways they are connected together. This form of relationalism "disavows any fixed absolute conception of space" (Murdoch, 1998: 358), being concerned instead with space as an arrangement of priorities (Star, 1995). Although it is necessary to be wary when generalizing on the basis of two case studies, I want to consider two specific effects of this relational topology, which I term "remote control" and "scalar abuse."

The topological ordering of modern conservation technologies enables them to remotely monitor and control a range of ecological "things." The fragmentation model *abstracts* ecological elements from their surrounding landscape and links them to distant other points, allowing "remote control," or "conservation at a distance" (Cooper, 1992). Returning to Figure 9.1 on page 242, the topological ordering of the fragmentation model results in "very distant points finding themselves connected to one another [points A and C] while others, that were once neighbours [A and B], come to be disconnected" (Murdoch, 1998: 360). The spatial practices of science, cartography, and technology behave as if ecological things—habitats, animals, landscapes—are separate from their local context. They decontextualize them in order to enroll them into remote political decisions. The case studies of Karvia and Vincent Drive, although vastly different in their details, essentially show how the topological ordering of the fragmentation model is achieved in practice (or, in terms of Figure 9.1, how points A and B are separated as A and C are brought closer together). The spatial politics of the fragmentation model provides a technology of remote control for conservation planners.

The topological character of the fragmentation model means that it can be reproduced at any scale. This is apparent in practice, as the fragmentation model is influential within conservation planning at a range of scales, from continents to hedgerows. This quality of the model permits the abuse of scale in its political implementation. In the case of the Finnish example, implementing the model at the European scale served to alienate the local foresters from conservation priorities. In the case of Vincent Drive, articulating the site in terms of its linear connectivity at the *city* scale (through the city network of corridors) rather than as a unique habitat patch at the *local* scale abstracted the site's ecology from its context (see Cowell, 2003, for a discussion of this type of scalar politics). The two examples highlight a crucial distinction here, though; the specific ends of the fragmentation model are not in any way predetermined by the model. In Karvia the model was being used to establish a continental conservation network, while in Birmingham the model was entirely complicit with the capitalist logic of development, which was justified primarily in terms of its benefit to the city (Urban Initiatives and Birmingham City Council, 1999). It is thus impossible to say that the fragmentation model is complicit with neoliberal development or that it is always a force for conservation. Its utility in the planning system lies in the fact that it can be either.[8] The linear relational ontology of the model permits remote control through the abuse of scale, but the specific ends of this process are not predetermined.[9] This quality makes it appealing to a range of interests in different contexts, forming what Maartin Hajer (1995) would term an attractive "storyline" for conservationists, planners, and prodevelopment interests alike in differing contexts.

The fragmentation model articulates a familiar form of institutional control. The function of the line is critical as a tool of administrative power—anyone living on the ground, who "knows" an area as an inhabitant, is aware that there is no line in practice. Institutional power is dependent upon the line, because the line simultaneously connects and separates. A number of authors have argued that more advanced conservation models (for example that of Pickett's, discussed above) founder upon the hard and fast linear divisions associated with property rights (Hurley et al., 2002). I would argue that the situation is not simply a reflection of property rights but rather a function of the line in delimiting spatial domains of institutional influence. As Doreen Massey argues (2005: 85), the "intrinsic relationality of the spatial is not just a matter of lines on a map; it is a cartography of power." In Deleuzian terms, conservation models operate as ordering machines; as they arrange humans and nonhumans in space, they simultaneously order domains of administrative control. The line can be reproduced at a range of scales without distorting the fundamental spatial and institutional segregation. The topological character of the fragmentation model means that it can be flexibly scaled to match political processes operating at different levels.

Planning is not ecologically naive: "The way ecological science achieves the effect of materiality in describing a 'territory with qualities' [is critical] because this achievement is essential in allowing economic and political strategies to effectively play out across that territory" (Robertson, 2004: 372). The effect of materiality is achieved through the spatial practices of conservation planning, but critically these practices cannot be cast in terms of simply left or right, prodevelopment or proconservation. The spatial politics of the fragmentation model remains influential, because it can be *either.* Although the regulatory apparatus of planning is closely entwined with capitalism, it is not reducible to purely economic logic. It is in the face of this paradox that it becomes appropriate to talk of a specifically *spatial* politics. In spite of the concerted efforts of academics to propose ways to integrate political scales with ecological scales in planning systems, the existing system is deeply embedded and extreme effort will be required not only to make new systems and scales of governance but also to somehow undo the ones that exist. Moving away from this form of conservation planning may be problematic. As Jonathan Murdoch (1998: 370) states, "Once one set of priorities has been imposed and enshrined in a given set of socio-materialities so they will tend to make the next 'round' of prioritising easier for those who were successful last time." The fragmentation model remains dominant in the face of contrary ecological evidence precisely because it speaks the spatial language of institutionalised power.

CONCLUSIONS: THE SPATIAL POLITICS OF SUSTAINABILITY

While conservation planning is a realm of activity that holds obvious importance for sustainability, I have argued that it is deeply inured to a series of regulatory realities that are both undertheorized and underestimated. At the beginning of this chapter I proposed that conservation models have a politics that needs to be questioned and that this politics is inherently spatial. These propositions were worked through with reference to the fragmentation model of conservation planning. The fragmentation model is a culturally and politically charged concept that is deeply embedded in the technical decision-making procedures of planning practice. Rather than judge its scientific merits, I have suggested that the key to understanding its political persistence lies in its ability to reproduce relational spaces that are amenable to regulatory control. As a politicized set of assumptions concerning humans and nature, the fragmentation model articulates a highly alluring arrangement of space in terms of *what* constitutes nature and *where* it should be.

The arguments of this chapter have in some ways been pessimistic, emphasizing inertia and persistence, but I want to end by exploring the potential space for radical change that some of these threads may open up. While the topological relational space of conservation planning *can* be produced at any scale, in practice it must always be embedded at a specific scale through spatial practices. Scale matters because it determines what we see—it is not merely a container but articulates political relations. The fragmentation model can be deployed to support the spatial requirements of development or conservation, but it is never politically neutral. We therefore need to be sensitive to the scales at which conservation models are deployed, as this has a direct effect on the ordering of space and the resultant priorities that are privileged. While the melange of science, policy, and practice that determines this process is complex, the politics of scale remains the language of radical change in conservation planning. It *delimits* and *delineates* the spatial politics of conservation planning, although the process of undoing existing institutionalized scales is dependent upon reforming a range of spatial practices.

Taking a step back from the problem, sensitivity is required concerning the spatial metaphors that inform our understanding of environmental planning. Returning to the logic of connection and insularity upon which fragmentation is based, parallels can be drawn with dominant circulatory ideologies of modernity concerning the city (Sennett, 1992). The logic of linear connectivity is intimately related to scientific discourses of circulation and health, and, in the case of the fragmentation model, the category of nature is bound up with the capitalist logic of healthy circulation. While politicizing circulation and metabolism in certain ways may "open up the theoretical and practical possibility for creating the environments we wish to inhabit" (Swyngedouw, 2004: 1), most obviously by offering a hybrid politics that promises to transcend dualisms between nature and humans, such circulatory tropes are animated by a potentially misleading anthropomorphism of landscape processes. The fragmentation model demonstrates exactly this, a point made beautifully by the cartoon in Figure 9.2, but the unquestioned dominance of circulatory tropes affects social scientific theory as well as scientific theory.

Rather than constantly seeking to impose new, more "sustainable," scales of planning, we could ask what might happen if we were to subvert the existing spatial language of planning from within. One solution is to engage in conservation planning at the most politically dominant scale of development. Some of the most successful conservation initiatives in highly modified landscapes occur at the scale of the building or the individual development. In terms of the dominant form of two-dimensional planning, technologies such as green and brown roofs provide habitat that occupies the *same* space as built development. Bat boxes and rough walls are also invisible in traditional two-dimensional space, occupying the vertical surfaces of human-made structures. It is possible to subvert the relational separation of human and natural spaces articulated by the fragmentation model by switching modes of representation. GIS certainly offers the technical tools with which to do this; however, new forms of planning classification would be needed that recognize the hybrid qualities of these "in-between," or three-dimensional, spaces.

Achieving such a progressive intermingling of humans, animals, and plants cohabiting in the same space transcends the boundaries of what is normally taken to constitute "conservation planning." On the level of lived experience, it prompts us to ask whether new relationships can be forged between humans and nonhumans based on cohabiting, or what has been termed "conviviality" (see Hinchliffe & Whatmore, 2006, for an exploration). On the institutional level, it suggests the possibility of reorganizing systems of administrative control to produce spaces that are less segregated. Rather than the either/or linear separation of the fragmentation model, this system would overlay nature conservation on to other land-use priorities. Each of these tasks implies rejecting the either/or dichotomy of development and conservation. Returning to the notion of "incomplete boundaries," perforated lines in space would require "incomplete institutions," but abandoning the hard-and-fast ontology of the line throws us back against the problems associated with legal jurisdiction, property rights, and the more general constraints of administrative control of which they are a part.

To return to the themes with which the chapter began, it is clear that the challenge of sustainable planning is a fraught political process. Not only is increasing pressure being exerted on space, but the institutions trusted with the responsibility to regulate that space are encouraged to proactively integrate different functions at a range of scales. Within this context, the spatial politics of planning is critical in shaping any radical agenda for change. Perhaps most importantly, it suggests that the process of shifting scales of governance is bound up with deeper issues such as our cultural relations with nature and the politico-ecological institutional expression of these relations in space. Understanding how they relate to one another and thus how they can be changed constitutes the major challenge to sustainable planning.

NOTES

1. Biological diversity was defined at the Rio Earth Conference as "the variability among living organisms from all sources including, *inter alia*, terrestrial, marine and other aquatic ecosystems and the ecological complexes of which they are part; this includes diversity within species, between species and of ecosystems" (Halpern, 1992, Article 2).
2. Following the Brundtland report, "sustainable development" was interpreted as

development that would seek "to meet the need of the present generation without compromising the ability of future generations to meet their own needs . . . [and] assuring the on-going productivity of exploitable natural resources and conserving all species of fauna and flora" (World Commission on Environment and Development, 1987, p. 43).

3. As some indication of this, searching Web of Knowledge (the premiere bibliographic service available to the British academic community) for journal articles taking conservation planning as their topic yielded some interesting results. The total number of articles on this topic increased from 18 in 1997 to 34 in 2001 and 95 in 2005. While far from an exhaustive analysis, this increase indicates the explosion of interest in this field over the past 10 years. The search was conducted on the expanded science, social science, and arts and humanities indexes (David & Zeitlyn, 1996).

4. It is perhaps worth noting that Edward Wilson, one of the authors, went on to become one of the high priests of "biodiversity" during the late 20th century (Takacs, 1996; Wilson, 1989).

5. The distinction between the political realm of conservation planning and the scientific realm of landscape ecology should be emphasized here. The focus of this chapter is what happens *in practice*. Rather than arguing that planners misapply the fragmentation model, I am suggesting that the model itself is amenable to misapplication.

6. In his excellent conceptual review of scale in ecology and geography, Sayre (2005) makes the distinction between the methodological and ontological moments of scale. The methodological moment concerns that age-old scientific conundrum that the scale at which a phenomenon is observed can determine the process that is observed (Phillips, 1999), with debate revolving around the choice of scale at which to study a phenomenon (Gregory et al., 2002). So, for example, a study of mean sea-level change in North England over the last thousand years may record a falling level due to isostatic rebound from the Pleistocene glaciation, while a study of sea-level change conducted at one place for an hour may record rising levels as the tide flows. At each scale sea-level change is driven by completely different processes. The ontological moment concerns scales that appear intrinsic to certain systems, such as watersheds, or the physical dimensions of an object. In the sciences, real processes operating at different scales are seen as independent of one another (Phillips, 1999), and although attention is given to how these scales interact and interlock through processes of energy and material transfer (Noss, 1992), "there is an 'incommensurable irreducibility' to nature when viewed and described at different levels" (Bauer, Veblen, & Winkler 1999: 3). Socioecological scale can be altered over time and space, however, so both social and ecological scales can be seen as produced. The advocates of dynamic scale assume that the ontological moment can be altered by the methodological—if we choose to plan at new scales, then reality will be shaped in their image.

7. The term "nonhuman" is used here as a term taken from Actor Network Theory to mean any entity that is not human—things, animals, knowledge, etc.

8. This is surely one of the reasons that the fragmentation model can be such a divisive issue. Conservationists defend it against ecological critique, because it is one of the few tools that has real influence within the planning system.

9. The topology of the fragmentation model prompts some reflection on how we conceptualize scale in the politics of sustainability. Notably, it prompts us to ask

whether the concept of scale has any relevance in understanding these processes, if it is so open to abuse. The classic topological argument concerning scale is based on the railway line analogy of Actor Network Theory (Latour, 1993: 117). The railway line, it is argued, is an example of something that is simultaneously local—there are always tracks "placed" somewhere—and yet supralocal, being coherent at the regional, national, or international levels. Scale simply becomes an effect of networks, depending on how long or short networks are. This is in stark opposition to ecological understandings of scale that see the methodological choice of scale as critical in determining the kinds of processes that are recorded. Returning to Figure 9.1, imagine an ecological survey of the imaginary space represented. If the survey is undertaken at the intrahabitat patch level, it will find homogeneity, whereas if the survey takes the landscape as its unit of study, then it will find heterogeneity. The key difference is that the railway line is only considered in terms of its own linear spatial and functional coherence. We are only offered to look at scales that are presented to us by it. The scientific view sees scale as a matter of methodological choice on behalf of the viewer that reveals different ontological realities. Thus, while scale can be described as an effect of networks, its status is not reducible to them.

REFERENCES

Adams, L., & Dove, L. (1989). *Wildlife reserves and corridors in the urban environment: A guide to ecological landscape planning and resource conservation.* Columbia, MD: National Institute for Urban Wildlife.

Adams, W. (1986). *Nature's place: Conservation sites and countryside change.* London: Allen & Unwin.

Barker, G. (1997). *A framework for the future: Green networks with multiple uses in and around towns and cities.* Peterborough, UK: English Nature.

Bauer, B., Ueblen, T., & Winkler, J. (1999). Old methodological sneakers: Fashion and function in a cross-training era. *Annals of the Association of American Geographies, 89,* 679–687.

Becker, S. (2002, February 25). Wildlife corridor monitors. *Grist: Environmental News and Commentary.* Accessed December 1, 2006, at *http://www.grist.org/comments/ha/2002/02/25/becker-wildlife/index.html.*

Beier, P., & Noss, R. (1998). Do habitat corridors provide connectivity? *Conservation Biology, 12,* 1241–1252.

Birmingham City Council. (1997). *Nature conservation strategy for Birmingham.* Birmingham, UK: Department of Planning and Architecture.

Boothby, J. (2000). An ecological focus for landscape planning. *Landscape Research, 25,* 281–289.

Boothby, J. (2004). Lines, boundaries and ontologies in planning: Addressing wildlife and landscape. *Planning, Practice and Research, 19,* 67–80.

Braun, B. (2005). Environmental issues: Writing a more-than-human urban geography. *Progress in Human Geography, 29,* 635–650.

Brenner, N. (2001). The limits to scale: Methodological reflections on scalar structuration. *Progress in Human Geography, 25,* 591–614.

Collinge, S. (1996). Ecological consequences of habitat fragmentation: Implications for landscape architecture and planning. *Landscape and Urban Planning, 36*, 59–77.

Collinge, S. (2001). Spatial ecology and biological conservation. *Biological Conservation, 100*, 1–2.

Cooper, B. (1992). Formal organisation as representation: Remote control, displacement and abbreviation. In M. Reed & M. Hughes (Eds.), *Rethinking organisation: New directions in organisational theory and analysis* (pp. 254–272). London: Sage.

Cowell, R. (2003). Substitution and scalar politics: Negotiating environmental compensation in Cardiff Bay. *Geoforum, 34*, 343–358.

David, M., & Zeitlyn, D. (1996). *What are they doing? Dilemmas in analyzing bibliographic searching: Cultural and technical networks in academic life*. Accessed April 14, 2001, at *http://www.socresonline.org.uk/socresonline/1/4/2.html*.

Dawson, D. (1994). *Are habitat corridors conduits for animals and plants in a fragmented landscape? A review of the scientific evidence*. Peterborough, UK: English Nature.

Desfor, G., & Keil, R. (2000). Every river tells a story: The Don River (Toronto) and the Los Angeles River (Los Angeles) as articulating landscapes. *Journal of Environmental Policy and Planning, 2*, 5–23.

Duane, T. (2000). *Shaping the sierra: Nature, culture and conflict in the changing West*. Berkeley, CA: University of California Press.

Dunning, J., Danielson, J., & Pulliam, H. (1992). Ecological processes that affect populations in complex landscapes. *Oikos, 65*, 159–175.

Eden, S. (1999). Business claims of legitimacy in the environmental debate. *Environment and Planning A, 31*, 1295–1310.

European Union. (1979). *Council directive 79/G09 EEC on the conservation of wild birds*. Brussels, Belgium: European Union.

European Union (1992). *Council Directive 92/43/EEC on the Conservation of Natural Habitats and of Wild Fauna and Flora*. Brussels, Belgium: European Union.

Evans, J. (2003). *Biodiversity conservation and brownfield sites: A scalar political ecology*. Unpublished PhD thesis, University of Birmingham, Birmingham, UK.

Evans, J. (2004). *Political ecology, scale and the reproduction of urban space: The case of Vincent Drive, Birmingham* (Working Paper no. 66). Birmingham, UK: School of Geography, Earth and Environmental Science, University of Birmingham.

Fabos, J., & Ryan, R. (2004). International greenway planning: An introduction. *Landscape and Urban Planning, 68*, 143–146.

Feldman, T., & Jonas, A. (2000). Sage scrub revolution?: Property rights, political fragmentation, and conservation planning in Southern California under the Federal Endangered Species Act. *Annals of the Association of American Geographers, 90*, 256–292.

FERN—The Forest and the European Union Resource Network. (2004). *Habitats directive and Natura 2000*. Accessed December 21, 2004, at *http://www.fern.org/pubs/briefs/Habitats%20Directive90200ct%2004.pdf*.

Forman, R. (1991). Landscape corridors: From theoretical foundations to public policy. In D. Saunders & R. Hobbs (Eds.), *Nature conservation 2: The role of corridors* (pp. 71–84). Chipping Norton, UK: Surrey Beatty & Sons.

Forman, R. (1995). *Land mosaics: The ecology of landscapes and regions*. Cambridge, UK: Cambridge University Press.

Gandy, M. (2002). *Concrete and clay: Re-working nature in New York City.* Cambridge, MA: MIT Press.

Glowka, L., Burhenne-Guilmin, F., & Synge, H. (1994). *A guide to the Convention on Biological Diversity.* Gland, Switzerland: IUCN—The World Conservation Union.

Gregory, K., Gurnell, A., & Petts, G. (2002). Restructuring physical geography. *Transactions of the Institute of British Geographers, 27,* 136–154.

Haila, Y. (1999a). Biodiversity and the divide between culture and nature. *Biodiversity Conservation, 8,* 165–181.

Haila, Y. (1999b). Socioecologies. *Ecography, 22,* 337–348.

Haila, Y. (2000). Beyond the nature–culture dualism. *Biology and Philosophy, 15,* 155–175.

Haila, Y. (2002). A conceptual genealogy of fragmentation research: From island biogeography to landscape ecology. *Ecological Applications, 12,* 321–334.

Hajer, M. (1995). *The politics of environmental discourse: Ecological modernisation and the policy process.* Oxford, UK: Clarendon Press.

Halpern, S. (1992). *United Nations Conference on Environment and Development: Process and Documentation.* Providence, RI: ACUNS.

Harvey, D. (1996). *Justice, Nature and the Geography of Difference.* Oxford, UK: Blackwell.

Haughton, G. (1999). Environmental justice an the sustainable city. *Journal of Planning Education and Research, 18,* 223–243.

Heynen, N., & Swyngedouw, E. (2003). Urban political ecology, justice and the politics of scale. *Antipode, 34,* 898–918.

Hiedanpää, J. (2002). European-wide conservation versus local well-being: the reception of the Natura 2000 reserve network in Karvia, SW-Finland. *Landscape and Urban Planning, 61,* 113–123.

Hinchliffe, S., & Whatmore, S. (2006). Living cities: Towards a politics of conviviality. *Science as Culture, 15,* 123–138.

Holling, C. (1973). Resilience and stability of ecological systems. *Annual Review of Ecological Systems, 4,* 1–23.

Hurley, J., Ginger, C., & Capen, D. (2002). Property concepts, ecological thought, and ecosystem management: A case of conservation policymaking in Vermont. *Society and Natural Resources, 15,* 295–312.

Jokinen, A. (2006). Standardization and entrainment in forest management. In Y. Haila & C. Dyke (Eds.), *How nature speaks: The dynamics of the human ecological condition* (pp. 198–217). Durham, NC: Duke University Press.

Keeley, B., & Tuttle, M. (1996). *Texas bats and bridges project.* Austin: Texas Department of Transportation.

Latour, B. (1993). *We have never been modern.* London: Harvester Wheatsheaf.

Lee, K. (1993). Greed, scale mismatch and learning. *Ecological Applications, 3,* 560–564.

Little, C. (1990). *Greenways for America.* Baltimore: Johns Hopkins University Press.

Mabey, R. (1999). *The unofficial countryside.* London: Pimlico.

MacArthur, R., & Wilson, E. (1967). *The theory of island biogeography.* Princeton, NJ: Princeton University Press.

Massey, D. (2005). *For space.* London: Sage.

McDonnell, M., & Pickett, S. (1993). *Humans as components of ecosystems.* New York: Springer Verlag.

McKibben, B. (1999). *The end of nature.* New York: Random House.

Miller, J., & Hobbs, R. (2002). Conservation where people live and work. *Conservation Biology, 16*, 330–337.

Mittermeier, R., Myers, N., & Mittermeier, C. (2000). *Hotspots: Earth's biologically richest and most endangered terrestrial ecoregions.* Chicago: University of Chicago Press.

Mol, A., & Law, J. (1994). Regions, networks and fluids: Anaemia and social topology. *Social Studies of Science, 26*, 641–671.

Murdoch, J. (1998). The spaces of actor network theory. *Geoforum, 29*, 357–374.

Murdoch, J., & Lowe, P. (2003). The preservationist paradox: Modernism, environmentalism and the politics of spatial division. *Transactions of the Institute of British Geographers, 28*, 318–332.

Nicholls, A., & Margules, C. (1991). The design of studies to demonstrate the biological importance of corridors. In D. Saunders & R. Hobbs (Eds.), *Nature conservation 2: The role of corridors* (pp. 49–61). Chipping Norton, UK: Surrey Beatty & Sons.

Noss, R. (1992). Issues of scale in conservation biology. In P. Fiedler & S. Jain (Eds.), *Conservation biology: The theory and practice of nature conservation, preservation, and management* (pp. 240–241). New York: Chapman and Hall.

Peck, S. (1998). *Planning for biodiversity: Issues and examples.* Washington, DC: Island Press.

Phillips, J. (1999). Methodology, scale and the field of dreams. *Annals of the Association of American Geographers, 89*, 754–760.

Pickett, S., & Cadenasso, M. (1995). Landscape ecology: Spatial heterogeneity in ecological systems. *Science, 269*, 331–334.

Pickett, S., Cadenasso, M., & Grove, J. (2003). Resilient cities: Meaning, models, and metaphor for integrating the ecological, socio-economic, and planning realms. *Landscape and Urban Planning, 69*, 369–384.

Pickett, S., Parker, V., & Fiedler, P. (1992). The new paradigm in ecology: Implications for conservation biology above the species level. In P. Fiedler & S. Jain (Eds.), *Conservation biology: The theory and practice of nature conservation, preservation and management* (pp. 65–88). London: Chapman & Hall.

Robertson, M. (2004). The neoliberalisation of ecosystem services: Wetland mitigation banking and problems in environmental governance. *Geoforum, 35*, 361–373.

Sayre, N. (2005). Ecological and geographical scale: Parallels and potential for integration. *Progress in Human Geography, 29*, 276–290.

Selman, P. (2002). Multi-function landscape plans: A missing link in sustainability planning? *Local Environment, 7*, 283–294.

Sennett, R. (1992). *Flesh and stone: The body and the city in Western civilisation.* London: Faber and Faber.

Slater, T. (2002). *Edgbaston: A history.* Chichester, UK: Phillimore & Co.

Small, E., Sadler, J., & Telfer, M. (2003). Carabid beetle assemblages on urban derelict sites in Birmingham, UK. *Journal of Insect Conservation, 6*, 233–246.

Smith, B. (1995). On drawing lines on a map. In A. Frank, W. Kuhn, & D. Mark (Eds.), *Spatial information theory: A theoretical basis for GIS* (pp. 475–484). Berlin: Springer.

Sneddon, C. (2002). Water conflicts and river basins: The contradictions of comanagement and scale in Northeast Thailand. *Society and Natural Resources, 15*, 725–741.

Spellerberg, I., & Gaywood, M. (1993). *Linear features: Linear habitats and wildlife corridors.* Peterborough, UK: English Nature.

Star, S. L. (1995). The politics of formal representations: Wizards, gurus, and organisational complexity. In S. L. Star (Ed.), *Ecologies of knowledge: Work and politics in science and technology* (pp. 89–118). New York: SUNY.

Swyngedouw, E. (1999). Modernity and hybridity: Nature, regeneracionismo, and the production of the Spanish waterscape, 1890–1930. *Annals of the Association of American Geographers, 89,* 443–465.

Swyngedouw, E. (2004). *Circulations and metabolisms: (Hybrid) natures and (cyborg) cities.* Accessed February 12, 2005, at *http://www.ru.nl/socgeo/n/colloquium/science. pdf.*

Takacs, D. (1996). *The idea of biodiversity: Philosophies of paradise.* Baltimore: John Hopkins University Press.

Thrift, N. (1996). *Spatial formations.* London: Thousand Oaks.

United Nations Center for Human Settlements. (1996). *An urbanizing world: Global report on human settlements.* Oxford, UK: Oxford University Press.

Urban Initiatives and Birmingham City Council. (1999). *Selly Oak development study.* Birmingham, UK: Birmingham City Council.

Whatmore, S. (2002). *Hybrid geographies.* London: Sage.

White, I., & Howe, J. (2003). Planning and the European Union Water Framework Directive. *Journal of Environmental Planning and Management, 46,* 621–631.

Whitfield, J. (2001, December 17). Conservationists patch it up: Urban wildlife may not use green corridors. *Nature News Service.* Accessed December 8, 2005, *http://www.innovations-report.com/html/reports/environment_sciences/report-6777.html.*

Whitmore, T. C. (1997). Tropical forest disturbance, disappearance and species loss. In W. Laurance & R. Bierregaard (Eds.), *Tropical forest remants: Ecology, management and conservation of fragmented communities* (pp. 3–12). Chicago: University of Chicago Press.

Wilson, E. (Ed.). (1989). *Biodiversity.* Washington, DC: National Academy Press.

Wilson, E. (1992). *The diversity of life.* Harmondsworth, UK: Penguin.

World Commission on Environment and Development, 1987. *Brundtland Report: Our Common Future.* Oxford, UK: Oxford University Press.

Worster, D. (1977). *Nature's economy: A history of ecological ideas.* New York: Sierra Club Books.

Yanes, M., Velasco, J., & Suárez, F. (1995). Permeability of roads and railways to vertebrates: The importance of culverts. *Biological Conservation, 71,* 217–222.

Zimmerer, K. (1994). The "new ecology" and human geography: The prospect and promise of integration. *Annals of the Association of American Geographers, 84,* 108–125.

Zimmerer, K. (2000). The reworking of conservation geographies: Non-equilibrium landscapes and nature–society hybrids. *Annals of the Association of American Geographers, 90,* 356–369.

CHAPTER 10

The Imperial Valley of California

Sustainability, Water, Agriculture, and Urban Growth

STEPHANIE PINCETL
BASIL KATZ

The concept of sustainable development as a way for humans to organize their activities into the future is yet another approach in the lineage of interest in, and concern about, humans' transformation of nature, from antiquity to the present (see, among many others, Coates, 1998; Glacken, 1967). It explicitly recognizes there are fundamental and inextricable linkages among economic, social, and environmental factors (Drummond & Marsden, 1999: 49), famously put forward in the Brundtland Commission report, *Our Common Future* (World Commission on Environment and Development, 1987).

The problem, however, has been how to understand and pursue this nexus. To date, the theory and practice of sustainable development has reached something of an impasse, because often this unity has been neglected (Drummond & Marsden, 1999). Moreover, sustainable development is an approach to conducting human affairs, not an analytical perspective.

As Sneddon (2000; among others such as Worster, 1993; Redclift, 1987; Escobar, 1995; Peet & Watts, 1996) has pointed out, sustainable *development* seems to have, as a concept and normative goal, become both the logical extension of efforts to reproduce development in a form more palatable to critics, and too vacuous and malleable to be of much use (Lélé, 1991, in Sneddon, 2000). Robinson (2004: 370) suggests that sustainable development is more attractive to government and business because it is a more managerial and incremental approach, similar to the Brundtland report. It does not fundamentally change the institutional structures that have arisen around production, and rarely engages in a strong critique of the methods and choices of production; rather, it introduces greater efficiencies. Sneddon (2000), therefore, proposes delinking sustainable development from sustainability, arguing that sustainability has the advantage of forcing a reference to specific geographic, temporal, and socioecological contexts. Context specificity forces the crucial questions of what exactly is being sustained, at what scale, by and for whom, and using what institutional mechanisms (2000: 525). Our approach seconds Bakker (2002: 774) who argues that the analysis of

> water (no less than other natural resources such as oil) plays a role in socioeconomic restructuring, and is both transformed by and constraining of political-economic choices and evolution. Analysis of the precise nature of this relationship is highly contingent, and can only be undertaken in specific historical–geographical contexts.

The Imperial Valley of California offers a complex case study to begin to apply this approach to sustainability. An irrigated desert in the southeastern portion of California, bordering on Mexico and Arizona, this region has historically been allocated the lion's share of water from the Colorado River, a watercourse that spans 7 states, 1,500 miles, and crosses an international border. It produces over $1 billion of agricultural products on 500,000 acres, and yet remains the poorest county (based on numerous metrics) in the state. In 2003 the Imperial Irrigation District—the water wholesaler and distributor in the Imperial Valley (and in the western United States in terms of volume of water)—was told by the federal government that it had to reduce its water use. This was the result of complex interacting socioenvironmental, political, and economic factors entrained in a cascade of changes, many of which are still unfolding.

In many ways, the developments in the Imperial Valley are a classic

case of "old" politics, where large powerful interests align around economic development strategies and enlist state agencies to support them.
Yet, as Bakker (2002) points out, such mobilizations are increasingly
supporting an organizational and/or institutional shift along a continuum of water management options toward a process of cohabitation,
competition, and eventual displacement of public policy decision-
making allocation principles based on social equity and economic self-
sufficiency, by market principles prioritizing economic efficiency and
allocation techniques (p. 769).

We highlight several issues in this case study, though there are many
others that could also be investigated. First, spatial scale. In this case, an
integrated interregional scale involving agriculture and urban areas,
coast and interior, is used to demonstrate the importance of examining
relevant trade-offs and how choices are made that may have unintended
long-term consequences when a larger landscape level is used. These include agriculture, cities, natural areas, community and economic development, equity and justice, and how they are linked together. We also
examine the roles of the institutions, structures, norms, and values that
define the rights, constraints, and power that have influenced and
shaped the development of region. These include the structures of political decision making, including multiple jurisdictions and single-purpose
agencies as well as private interests. The environmental impacts of agriculture in this region and the question of how the Colorado River water
should or will be divided into the future form another aspect of our
analysis. Finally, this case study illustrates how introducing new allocation rules—in our case, a water transfer—engenders a new hydrosocial
contract between users and their environment and a new exchange value
to the resource.

SUSTAINABILITY AND SUSTAINABLE DEVELOPMENT

There has been a great deal written about sustainable development and
sustainability. It is not our goal to review the literature but simply to
suggest that the linkages between environmental degradation and the institutions, structures, and norms that constitute and create our current
social organization(s) are the critical ones to explore and tend to have
been neglected.

Sustainable development, paradoxically, often lacks specificity relative to how environments are materially transformed for the mainte-

nance of current lifestyles, and through what regulatory mechanisms. Moreover, too often sustainable development—and sustainability studies— also neglects the powerful effects of the transformation of the environment on human society itself.

> In our complex dealings with the physical world, we find it very difficult to recognize all the products of our own activities. We recognize some of the products, and call others by-products; but the slagheap is as real a product as the coal, just as the river stinking with sewage and detergent is as much our product as the reservoir. (Williams, 1980: 83)

In the Imperial Valley, the biophysical effects of the human transformation of nature are a significant cause of high incidence of asthma and a toxic sea and are now also significant constraints on policy options, as we will discuss.

In a country-by-country discussion of how governments of different highly developed countries have engaged with the idea of sustainable development, Lafferty and Meadowcroft, in their edited volume on *Implementing Sustainable Development, Strategies and Initiatives in High Consumption Societies* (2000), provide a useful overview of how countries have operationalized the concept. The volume's chapters illustrate differing regulatory styles, policy mixes, and sets of preferred instruments (tools), but relative to our concern about examining the ways in which advanced societies engage with the material basis of their reproduction and the rules regulating that engagement, the regulatory and policy descriptions fall short. For example, are there conventions and subventions that are important in structuring current agricultural practices? At what scale? Who owns the primary resources, such as agricultural land and timber lands, and how are access and exploitation regulated? Where does water come from? In what condition are the soils? To know whether policies are creating more sustainable conditions, greater investigation into these relations—highlighting the modes of regulation— are useful.

Timothy Beatley has also written a great deal about sustainable communities, noting that such communities acknowledge fundamental ecological limits (1995: 383). A sustainable community "is a place that seeks to minimize the extent of the urban 'footprint' and strives to keep to a minimum the conversion of natural open lands to urban and developed uses. . . Sustainable communities are, then, places that exhibit a compact urban form" (Beatley, 1995: 384). This includes reducing the environmental impact of buildings and taking into account regional environmental considerations, most particularly the preservation of ecosystems (Zimmerman, 2001). Yet again, these prescriptions tell us little about how it is rural spaces have come to be targets of land development that threatens biodiversity, nor where and how communities will provision themselves. Does it mean communities should exploit their own hinterlands for their needs? Does this not risk a collision course with biodiversity? We try to acknowledge and investigate what a sustainability perspective means relative to these questions in our case study, where clearly there are important linkages between the exploitation of the Imperial Valley and the viability of proximate urban centers that are structured by rules.

Accounts of sustainability like Beatley's have often been criticized for being predominantly concerned with the natural environment—calls for urban sustainability are justified because of environmental impacts. Indeed, sustainability has been criticized for starting with a broad agenda but moving quickly to forms of ecological or environmental determinism (Perkins & Thorns, 2001). But perhaps the problem is that sustainability is not sufficiently engaged with unraveling how we organize ourselves to appropriate the natural environment for our material basis for life. In other words, the appropriation of natural resources and their transformation into material goods (including water and food) are effectuated through institutions and institutional arrangements that are organized in certain patterns. Each regime of accumulation entails a mode of nature appropriation that produces a dynamic resource landscape in its own image (Bakker, 2002), and each has socioeconomic ramifications. For there to be greater sustainability, then, such linkages, patterns, and consequences need to be better understood.

Klug (2002), for example, points out how the unarticulated assumptions of property rights, embedded in the interstices of law, threatens ecosystem management and aquatic resources management. The deeply structuring rules of a mode of production constrain change in environmental management. Klug explains how, even with official recognition of ecosystem management, the property rights regime in the United States may profoundly constrain the sustainable management of renewable natural resources (p. 694). In the Imperial case, what is emerging is a political strategy by farmers to create property rights in the Colorado River water they use to irrigate crops. If such a change were to occur (and the current Bush administration is already attempting to create such rights in water elsewhere in California), the new regulatory structure could dramatically affect the future of the Valley and beyond.

There are, as Goodman, Sorj, and Wilkinson (1987) explain, tremendously powerful forces at work attempting to fundamentally reconfigure agricultural production. This involves not only the reworking of the basic genetic material of seeds (and appropriation of that genetic information) but also reducing the amount of primary resources in food through remanufacturing and reducing human inputs throughout. A profound reconfiguration of the relationship to nature is being undertaken, including greater manipulation of natural processes and appropriation of knowledge. Though there are irreducible biophysical inputs that must come from somewhere (such as water and land), the process of raising crops is being transformed, and the role of the farmer relative to nature and to production, as well. Understanding the relationship between these changes, and what forces are driving them, informs our understanding of the evolution of property rights in natural resources and particular social–natural articulations.

Donald Worster (1993: 143) writes that sustainability came about because the true way to curb the destructive path of Western capitalism was that "there must be limits to growth," and this was unacceptable. While it is beyond the scope of this article to assess whether Western capitalism is compatible with limited growth, ecological economists have been developing alternative economic valuations that go beyond simple material growth based on increased production. Over the course of the 20th century, and starting even with the biophysical economics of the 18th-century physiocrats (Cleveland, 1987) some observers argued that economic processes should also be conceptualized in terms usually used to describe processes in nature. Ecological economists include in their analysis the nonreplicable contribution of energy and materials and the depletion of thermodynamic sources, creating an expanded balance sheet that includes time and space dimensions: the relationship between dynamic human economic systems and larger dynamic but normally slower-changing ecological systems (Costanza & Daly, 1992; Costanza, Farber, & Maxwell, 1989; Costanza, Cumberland, Daly, Goodland, & Norgaard, 1997; Cleveland, Stern, & Costanza, 2001; Cleveland, 1991). The modes of production and the ways in which the appropriation of nature takes place affect the pace and scale of use of nonrenewable resources and human well-being. Ecological economics is adding an additional dimension to the toolkit for evaluating sustainability or unsustainability.

Sustainability approaches and practices, therefore, are dependent on institutions and practices that structure the socioenvironment. Regulation theory offers a framework for unraveling human social organization, including the mode of production that underlies the appropriation of nature. As Drummond and Marsden (1999: 40) explain,

> Modes of social regulation are constituted in the institutions, structures, norms and values which cede coherence to particular phases of capitalist development. . . . By defining rights, constraints and power, which in turn influence the ways in which real causal mechanisms are expressed in practice, they serve to license and to some extent direct, the nature of development."

Though regulation theory has primarily been applied to questions of capitalist accumulation and its impacts on society, Drummond and Marsden argue it also applies to the structures that shape human relaions with the environment. A regulationist approach highlights the structures that shape the politics of ecological appropriation.

We attempt, using a regulationist approach combined with an emphasis on political ecology, to examine the complex situation of water in southern California. We aim for a careful examination of the institutional framework for decision making—the context of regulation—to develop an interpretive understanding of what has shaped the current water transfer. We identify the causal mechanisms and how they work (Sayer, 2000) to ask what sustainability could look like taking into account specific geographical, temporal, and socioecological contexts. This will allow us to explore what exactly is being sustained, at what scale, by and for whom, and using what institutional mechanisms.

THE CASE STUDY

In California, cities and agriculture receive water from federally subsidized water projects. We begin our analysis recognizing that the long-term future (sustainability) of these water projects is in question due to the potential impacts of climate change on the hydrologic cycle as well as increased water-intensive urbanization in several of the states receiving water from the Colorado River. Agriculture in California receives the majority of developed water—approximately 80%. With increased urban growth in the state, there has been pressure to transfer some of the

water from agriculture to urban areas. In 2003 such a transfer was en-
acted, wherein the Imperial Valley Irrigation District entered into an
agreement to transfer a portion of its water to the San Diego County
Water Authority.

We first provide background on urban–rural relationships in water
development in the state and then discuss the Imperial Valley case. We
then turn to a discussion of institutions, spatial scale, and interregional
relations and conclude by discussing how the new allocation rules—in
our case, a water transfer—engenders a new hydrosocial contract between users and their environment and a new exchange value to the resource.

WATER IN CALIFORNIA AND ITS TRANSFER
FROM AGRICULTURE TO URBAN AREAS

Water development and agriculture in California grew up together, inextricably bound, for one would not have existed without the other.
Historically, water development in the West was predicated on providing
inexpensive water for agriculture, with urban areas paying the great
bulk of the capital costs. Recently, as the Imperial Valley water transfer
demonstrates, this relationship is under challenge and has begun to
change.

Background on Rural–Urban Relations
in Water Development in California

There has been a great deal written on California's water development,
and therefore we will not review this history (Worster, 1985; Reisner,
1986; Gottlieb & FitzSimmons, 1991; Hundley, 1992). Instead, we will
concentrate on events pertaining to the Imperial Valley and its development. It is important to note, however, that federal water development in
the West constitutes a kind of regulatory regime whose component parts
are fundamentally intertwined—rule changes for one part of the system
have implications for all other water projects that were developed under
the 1902 Reclamation Act. For example, currently the developed water
is considered a common good regulated by the federal government. If
water rights were to be transferred to farmers in the Imperial Valley, it
would have a cascade effect on all federally subsidized water provided to
farmers. The water right would become privatized. It is also important

to understand that both the federal and California state water projects have provided agriculture in California with inexpensive water while urban areas paid considerably more—and subsidized water development—based on the historic premise of the importance of agricultural production to the nation. Finally, this water transfer can be seen as a transformation of the Progressive Era water regime toward a new and yet emerging regulatory regime.

By the 1980s the conventional appropriation of water began to be challenged, triggered by proposals to increase water conveyance to southern California. Environmental organizations were concerned about the environmental effects of water transfers on the places of origin and the Bay Delta system, about the environmental impacts of agriculture itself, and also by concern that increased water would fuel further urban sprawl.

Water in California is delivered locally by local water districts, and the institutional relationships are between the federal government and the local water districts for federally developed water systems. For the water conveyed in the state-financed State Water Project, the institutional relationships are between local water districts and the state Department of Water Resources. Southern California receives water from the federal Colorado River delivery system—up to 12% of California's developed water; the rest comes from the State Water Project (SWP). (The city of Los Angeles receives water from the Owens Valley through its own conveyance system.) The water is then provided to several irrigation districts—including the Imperial Irrigation District—as well as to the Southern California Metropolitan Water District (MWD). MWD is a water wholesaler to 95% of the South Coast region. The agency includes 14 cities, 12 municipal water districts, and the San Diego County Water Authority (SDCWA). The SDCWA provides water to the second-largest metropolitan area in the state. The MWD and the SDCWA are key players in the SWP and the Colorado River delivery system because they draw from both—thus implicitly having a tie-in to the federal water project in California, the Central Valley Project (CVP). As such, the MWD and SDCWA have ended up as strategic players in the entire state, and indeed the whole West. What they do has a kind of "cascade" effect on water use and water development over very wide spaces.

Water agencies in California come in many different sizes and governance arrangements, including one-person, one-vote; property-weighted voting; and appointed boards. The Imperial Irrigation District (IID) Board is a district-elected board.

FIGURE 10.1. Southern California counties, showing the location of Imperial County.

Introduction to the Imperial Valley Case

Before the creation of the IID there were private efforts to harness the Colorado River for irrigation in the Imperial Valley. These efforts have also been described by others (Hundley, 1992; Gottlieb & FitzSimmons, 1991; Worster, 1985; Cory, 1915), and we will not recount the Valley's early history except to explain that by 1901 private efforts had succeeded in diverting water to the Imperial Valley and by 1905 there were already 120,000 acres of land under production. However, in 1905 a swollen Colorado River washed away the primitive irrigation canals and control gates, and water came flooding into the Salton Sink, refilling the northerly portion of a once prehistoric seabed. The Salton Sea was then reborn. The capital costs of irrigation infrastructure led to the creation of the Imperial Irrigation District in 1911 by Imperial Valley businessmen, who submitted the question of the creation of the district to the local voters (permissible under California law). Its governance was established as a five-member elected board, each member representing a division within the IID's service area. The new IID began rebuilding the water delivery system and installing a farm drainage system (tiles underground) capable of carrying away excess water from the heavy clay soils, and then it lobbied the federal government to build dams on the Colorado River to regulate its flow. The IID was successful in its efforts, and once federally subsidized water was available—through the construction of dams and canals on the Colorado River—the IID was the agency in charge of its distribution. The district also went into the energy sales and distribution business.

Salinity is an especially serious threat along the lower Colorado River, where the concentration of salts has increased as the volume of water decreases because of evaporation losses at reservoirs and intense irrigation practices upstream. Farmers must irrigate fewer acres with the same volume of water, switch to more salt-tolerant crops, install an expensive system of tile drains, obtain more water to produce the same crops, or adopt some combination of these options. Here we find an example of how humanly altered nature in turn affects human practices. Nature and humans are indeed intertwined.

Soils are reputed to be varied in the Imperial Valley, though there is wide agreement that there is a "pan" that alluvial drainage has been depositing over geologic time. Besides the runoff from the surrounding mountains, periodic flooding when the Colorado River changed its course over millennia and then alluvium drying have both contributed to the deposition of materials. Thus, there is no soil homogeneity. Interestingly, there is no available U.S. Natural Resources Conservation Service soil survey for the Imperial Valley.

The Imperial Valley has spent millions of dollars in a struggle to control salts that, barring some unexpected technological breakthrough or infusion of new water, will inevitably be lost (Hundley, 1992: 374). To neutralize the problem of salinity, farmers must flood their fields so as to drain off the salt, contributing to high water use in the valley. The water, highly laden with saline content, fertilizer, pesticide, and herbicide, is then drained to the Salton Sea, whose salinity levels and water quality have risen to near toxic levels for fish and fowl. Salinity levels have reached 44,000 milligrams per liter, approximately 25% saltier than seawater. Bird and fish kills are commonplace.

Once the federal government became involved in the development and allocation of Colorado River water, the question was how to divide it as it flows through seven states and into Mexico, emptying into the Sea of Cortez. The solution developed at the time was to divide the states into upper basin states and lower basin states. Water flows for the river were estimated to be about 17.5 million acre-feet a year (MAF). Since the water originated in the upper basin, any arrangement required the upper basin states to guarantee water for the lower basin states. Each basin was to receive 7.5 MAF a year, and Mexico was to receive 1.5 MAF leaving 1 MAF of surplus. The 1922 Colorado River allocation figures were based on unusually high river flows (17.5 MAF a year), today we know that the river flows closer to 12–14 MAF a year, leaving potentially significant shortfalls in promised water allocations.

The 1922 Colorado River Compact allocated 4.4 MAF to California. However, until very recently, other states along the river did not use their entire allocation, and the river's flow during the past half-century was unusually high. California negotiated to receive the surplus under the compact. Arizona built the Central Arizona Project (completed in 1998), and started taking its allocation to supply new urban growth, threatening a reduction in California's water surplus. In turn, this meant that the MWD would start getting reductions, and the MWD started to fear that it would not be able to deliver water for all the new growth occurring. California had been using 5.2 MAF and the IID 3.1 MAF, about 60% of California's allocation. The MWD, its member cities, and 2 Imperial Valley area water districts received about 2.1 MAF. Some entity had to give up water to get consumption down to the allocated 4.4 MAF.

There are approximately 500,000 acres farmed in the Imperial Valley, producing over $1 billion of food each year (see Table 10.1).

Approximately 30% of the farms in Imperial County are larger than 1,000 acres. The county ranks second in the state for average farm income, and the average annual earnings for each farm totaled $383,172 (*http://www.ccbres.sdsu.edu/data/indicators_of_the_month/pdf/2004/ article_december.pdf*).

Alfalfa ranked first in terms of 2003 farm acreage, accounting for 177,964 acres; next in rank were, successively, bermuda grass (a turf grass for golf courses), sudan grass (a feed grass), wheat, and sugar beets. These top five crops account for over half of the irrigated acreage in the valley. Vegetables represent the biggest commodity, accounting for

TABLE 10.1. Size and Concentration of Farm Ownership

Acreage	Number of farms
1–9	54
10–49	64
50–179	94
180–499	89
500–999	78
1000+	158
Total	537

Note. Data from U.S. Department of Agriculture, National Agricultural Statistical Service, 2002, Census of Agriculture.

over one-third of production. The Imperial Valley also happens to be a large producer of beef stock, boasting a new packing plant.

Ironically, Imperial County is also the poorest county in the state, due to its having the highest unemployment rate (over 25%, according to 2000 U.S. Census figures) and low wages. It also has a serious hepatitis C epidemic, air pollution among the worst in the country, 16% of births to teen mothers, fully 50% of children 0–4 living in poverty, an infant death rate almost three times that of California as a whole, one in every three inhabitants having no medical insurance, and fewer than one primary-care physician for every 3,500 persons. What explains this seeming paradox is the mode of agricultural production— how farming is organizationally structured. Agriculture in the Imperial Valley is characterized by a concentration of largely absentee owners (as our chart shows), the employment of tenant farmers or farm managers, and the widespread use of seasonal labor. This form of agriculture is highly articulated with industrial supply chains at both ends: inputs—including chemical fertilizers, pesticides, herbicides, and large-scale machinery—and output (products) feed into a highly organized manufacturing and distribution network. This system is supported by complex financial arrangements and is highly regimented in its activity. Large-scale agriculture in California has powerfully resisted better working conditions and pay for day laborers and has exercised strong political influence at the state and federal levels over these issues for decades (see, among others, Rudy, 1995; Pincetl, 1999; McWilliams, 1948/1968, 1949; Goldschmidt, 1946). Agriculture as it is currently practiced is an extractive activity that is greatly dependent on high energy and nonrenewable inputs. It is also highly productive. Organic production is also beginning to take hold in the Valley, though figures for acreages or sales are not yet available. One of the emerging questions is whether this might provide a direction for a more sustainable exploitation of land and water due to less use of chemical inputs and more careful irrigation practices.

THE DECISION-MAKING PROCESS LEADING TO THE WATER TRANSFER

In the Imperial Valley

Under current water law, to receive water from the Bureau of Reclamation or from the state it is necessary to demonstrate a "need" (Water

Code 100–100.5). With diminishing water availability from the Colorado River, demonstrating "need" has become progressively more important. During the early 1980s, two large landowners in the Elmore family, second-generation landowners in the district, challenged the IID's water use. Due to district-sponsored drainage policies (to leach salts), the level of the Salton Sea kept rising, flooding the Elmores' adjacent agricultural lands. The Elmores initiated action against the district before the state Water Resources Control Board (WRCB) and the state Department of Water Resources, accusing the IID of wasting water.

Subsequent investigations by the WRCB and the DWR (State Water Resources Act, 1945) found that the IID could conserve 438,000 acre-feet of water a year by lining its earthen canals (which leaked), by improving imprecise measuring systems and inefficient distribution systems, and by reducing spillage during farm use. The practice of flooding fields, using gravity irrigation—and exaggerated salt leaching—and growing high water-use crops all did not meet best-management practices. Thus, the existing mode of production—the organization of how water was delivered and used (regulated by the IID and federal rules)— had itself led to waste. Moreover, DWR also found a huge amount of nonrecoverable losses attributable to evaporation. Such a finding implicitly jeopardized agriculture's share of Colorado River water and called into question IID's water management authority.

Initially rejected by the IID and denied as a practicable possibility by a representative of the MWD in the DWR, a proposal was put forward by the Elmores that water conserved within the IID could be exchanged with the MWD for financial help with implementing conservation methods (Rudy, 1995). Given the impending completion of the Central Arizona Project, thereby reducing California's access to surplus (as well as the defeat by voters in 1982 of the Peripheral Canal, the other possible source of water for southern California), water trading began to be seen as a viable option. Water trading was also being advocated by the Environmental Defense Fund (EDF), a nonprofit environmental organization that had been deeply involved in the fight to defeat the Peripheral Canal (Rudy, 1995: 436–437).

DWR's finding was a foundational first step for challenging agricultural water use in the Imperial Valley. The groundwork had been laid for the IID to be able to *sell* agricultural water to growing urban and metropolitan areas in southern California as a more efficient use of water. This would also ensure that cities would continue to have sufficient water for their needs despite California's having to reduce its share of Colorado

River water. The selling of water to urban areas represents the introduction of the logic of the market to water resources management and allocation, a commercialization of the resources. It carries with it the application of private sector criteria such as efficiency and pricing.

The IID, however, asserted that it could not shoulder the cost of putting conservation measures in place. The MWD, in response, proposed a water marketing scheme (inspired by the EDF) that would pay for conservation measures and liberate hundreds of thousands of acre-feet of water for the southland (Gottlieb & FitzSimmons, 1991: 81–82). After much infighting and intense negotiations, an initial agreement was signed in 1988 in which the MWD would pay for certain conservation projects in exchange for water. This agreement, however, crumbled after the Coachella Valley Water Authority (CVWA), just to the north of the Imperial Valley, declared its water rights to be violated by the accord.

Historically, the CVWA had invested more money and greater effort into conservation. Even during the 1940s, "[its] farmers received their water, not through earthen or even concrete lined canals as in the Imperial Valley, but through a system of underground pipelines which eliminated losses due to both evaporation and seepage. Moreover, unlike IID's reliance on estimated flow, water deliveries in CVWD were measured through meters installed on every farm" (Waller, 1994: 24). Investments in efficiency continued through the years, including water deliveries allocated by a central computer, drip irrigation on 37% of irrigated acreage, and wastewater reclamation for urban irrigation (Waller, 1994: 24). Yet, the negotiations between the IID, the MWD, and indirectly the San Diego County Water Authority ushered in the new conceptions of the marketability of water to southern California, a concept that had already gained momentum in other parts of the state. This represented potential change in regulatory rules for federal water.

In 1996, then U.S. Interior Secretary Bruce Babbitt told California to come up with a plan to bring its Colorado River water draw back to the legal limit of 4.4 MAF a year within 20 years, because Arizona was beginning to draw its full allotted water (through the Central Arizona Project), as was Nevada. The MWD and other water agencies came up with a plan to reduce their dependence on the Colorado River, relying on a series of complex water transfers and water storage plans. Just before leaving office, Secretary Babbitt issued an order requiring all parties to agree to the 4.4 plan by December 31, 2002—called the Quantification Settlement Agreement (QSA). If the QSA were not agreed to by then, the temporary permission to exceed the 4.4 MAF threshold would be with-drawn (Fine, 2002).

By 1997, following federal actions to reduce California's allocation, the Colorado River Board of California drafted an agreement, titled California's Colorado River Water Use Plan, calling for the IID to sell part of its Colorado River allocation (200,000 acre-feet) to San Diego in support of the plan's goal of using Colorado River water more efficiently (lining canals with concrete, as in the Coachella Valley, for instance). The water would be conveyed through MWD facilities. When this deal was first proposed, the San Diego County Water Authority (SDCWA) objected to the price that the MWD said it would charge to transport the water through its facilities. That issue was resolved when the state agreed to pay $200 million to make up the difference.

However, in 2001 environmentalists raised concerns about the effects of the diversion on the shrinking Salton Sea and the species of fish and birds that rely on that sea. In December 2002, negotiations for the water transfer broke down as the district claimed it was being railroaded and objected to specific parts of the deal. The federal government then carried through on a threat to reduce water supplies: in the spring of 2003 U.S. Interior Secretary Gale Norton cut Colorado River water to Imperial County by 11%. Intense last-minute negotiations at the end of 2003 led to the approval of the QSA, transferring water from the IID to San Diego County Water Authority and to the Coachella Valley Water District. The federal government, pressured by environmentalists, instructed the IID that before it could sell the water to San Diego it must somehow mitigate the impact of the loss of drainage water ("wasted water") into the Salton Sea that was to go to San Diego. A polluted man-made sea had become an important bird habitat, thereby constraining decision making about water allocations—a new nature had arisen with its own new requirements.

The Quantification Settlement Agreement was also accompanied by three bills, establishing state policy with respect to the Salton Sea (State Bill 277, sponsored by Ducheny; SB 317, Kuehl; and SB 654, Machado). These bills stated the intention of the legislature that the state undertake restoration of the Salton Sea ecosystem and permanent protection of the fish and wildlife dependent upon that system. The legislation enables the Department of Water Resources to purchase Colorado River water from the IID and to sell it to MWD under specified terms. Proceeds from the sale of the water—estimated at perhaps $300 million—are to go to the Salton Sea Restoration Fund (State of California, Department of Water

Resources, 2004), and the agreement requires the transfer of up to 150,000 acre-feet of water to the Salton Sea to maintain its level. With over $300 million allocated to Salton Sea restoration (but no plan), environmentalists signed on to the IID transfer of water to San Diego.

Yet, it is the agricultural water-use inefficiency that provides the water sustaining the Salton Sea. The most significant impacts of a water transfer that generated water by improving efficiencies would be manifested at the Salton Sea. Agricultural drainage and administrative spills account for more than 85% of the sea's total annual inflow of 1.36 MAF. Agricultural drainage is a combination of tailwater (from field runoff, carrying high concentrations of fertilizers and pesticides) and tilewater (water that has leached through the field, with higher salinity and selenium loads, and carried to the sea through underground tiles). On-farm conservation measures would tend to reduce tailwater, decreasing nutrient loading to the Salton Sea, while increasing the relative concentrations of salts and selenium. Improvements to IID's delivery system would reduce the volume of the best-quality water reaching the sea. Fallowing would have the least negative impact on the sea, but taking agricultural land out of production directly and indirectly decreases local agricultural employment and could accelerate land conversion to urban uses. Also, fallowing does not address the SWRCB's findings of inefficient use (Pacific Institute, 2003) and poses air quality problems from dust from fallowed fields. The Imperial Valley is already out of the federal Clean Air Act air quality attainment. The planned lining of the All American canal will have consequences both for Mexican agricultural water supplies and for wetlands in Mexico, both dependent on underground water seepage from the unlined canal. Less water in Mexico will mean fewer jobs. Moreover, water diversions have devastated the Colorado River delta, which once covered 1,930,000 acres and is now sustained by less than 100,000 acre-feet of water and some underground seepage from irrigation practices.[1] The Colorado River water's regulatory regime, established by the compact, has entrained a complex set of consequences. It enabled farming on both sides of the border, perpetuated a new water body, created new wetlands, and nearly dessicated abundant wetlands at the Colorado River delta. These were enabled due to the rules governing water allocation, sale, and use, relations between local irrigation districts and the federal government, and the structure of agriculture.

Water transfers from the Imperial Valley involve a combination of agricultural land fallowing—more in the initial years—and the eventual lining of the All-American and Coachella canals, which will yield up to 100,000 acre-feet per year to the Metropolitan Water District. Over all, the transfer will provide the San Diego region with up to an additional 21.5 million acre-feet of water over the 110-year life of the agreement (San Diego County Water Authority, 2004). The transfer will also provide up to 150,000 acre-feet a year over 15 years to the Salton Sea (Revised Fourth Amendment Agreement, 2003).

The water transfer from the Imperial Irrigation District to San Diego has created a series of policy trade-offs: fallowing land is the most immediately available way to liberate water for the transfer to San Diego, does not decrease the "wastage" of water in agriculture, but ensures that the Salton Sea continues to receive the water it needs to be sustained. Fallowing increases air pollution in the Valley and reduces employment. The planned lining of irrigation canals and other water conservation measures will reduce water seepage, upon which a portion of Mexican agriculture depends, as do also valuable wetlands. So, though still emergent, the water transfer creates a new regulatory regime for water.

There is ample evidence that there are promising agricultural practices that will conserve water as well as counteract salinity, improve water flow through the soil, and improve soil moisture. Drip irrigation is used on less than 5% of the irrigated land in the Imperial Valley, mostly on vegetable crops and citrus (*California Agriculture*, 1997). It can, in the right situations, be a good choice in lands prone to salinization, as is shown in the Coachella Valley, just to the north of the Imperial Valley and still in the same geological formation.

The water transfer also comes about at a time when there are other shifts in the Valley. The county's population is growing rapidly, having expanded 30% from 109,303 in 1990 to 142,361 in 2000 (2000 U.S. Census). Land development is proceeding apace, and large developers such as Kaufman and Broad, on the housing side, and Wal-Mart, on the retail side, have been very active, with 3 Wal-Marts slated to be built and many thousands of housing units approved. Cities in the county, like the county itself, are poor, and they will receive none of the proceeds of the water sales, as the IID is an independent agency. Opportunities for growth are seen as a way to increase sales taxes and to become more prosperous. Since there is a lot of water in Imperial County, cities have little incentive to consider the impact of growth on that resource and consequently do not do so.

Water speculators who thought they could make a long "play" on the lucrative conversion of irrigation water stepped in during the initial discussions about water transfers during the late 1970s. First was the Fort Worth-based Bass family, who failed in an attempt to sell the water they received on their farmland to San Diego County directly. They were paying $15.50 a cubic acre and sought to sell the water to San Diego for up to $400 a cubic acre. Their land was then bought for $200 million in 1987 by U.S. Filter, a subsidiary of French conglomerate Vivendi Environment. With this purchase, U.S. Filter owned 42,000 acres of valley farmland, about 10% of the total under irrigation (Tempest, 2002). In February 2004, U.S. Filter, by then a unit of Veolia Environnement, sold its holdings to the Imperial Irrigation District for $77.3 million in cash (Imperial Irrigation District, 2004; Veolia Environment, 2006). This land is now being fallowed.

The agricultural sector in the Imperial Valley directly employs 8,438 workers a year, of which 5,044 are employed less than 150 days a year (2002 Census of Agriculture, County Data, p. 262); many of these are migrant workers who come over the border every day. Wages are low. The agricultural economic base of the county has not created prosperity (Barclay, Schmidt, & Hill, 1980; Worster, 1985; Rudy, 1995). The key question for the Imperial Valley is the future of agriculture relative to other forms of land use such as residential and commercial development that would potentially yield greater revenues directly to the cities and the county, and probably higher wages. In essence there would seem to be parallel gains to be made both by the IID for funds derived from the water sales from fallowed land and by the cities and counties for funds that would be generated by urban growth. Each entity gains from the relative decline in agriculture.

Yet, agriculture exhibits unique characteristics (Guthman, 2004: 63–64). The first is that food systems are fundamentally dependent on biophysical production to the extent that much of the value is created by biological processes. The rhythmicality and seasonality of many biological processes limit (still) the extent to which food production can be controlled or sped up and where food can be produced. From this perspective, the Imperial Valley offers exceptional conditions for growing, so long as water and drainage are available. Crops can be grown year-round, and a number of them can be harvested three to four times annually in the Imperial Valley. Finally, land is the major medium of agricultural production. It is unique—and unsubstitutable in the absolute sense. Once it is developed for other uses, the costs of reconversion are very high—if it can be done at all.

Goodman et al. (1987), in their highly thoughtful and sobering examination of the future of agriculture, do point out the increasing tendency to reduce the contribution of agricultural products to a smaller and smaller part of value added in favor of the production of synthetic fibers and food substitutes from hydrocarbons (p. 8). The relationship of food manufacturing to agricultural materials becomes purely instrumental, and the push is toward interchangeability—reducing them to an input. This, though, requires other types of inputs, including energy—not only in the remanufacturing of the initial input but also in transportation and processing. In an era of uncertainty about oil—whether its availability, its price, its quantity, or its climate-changing impacts—intensification of substitutability remains uncertain. On the production side, land and water are seen as mere inputs, not as important natural phenomena put to service in a particular manner by human social, political, and economic organization that, if depleted, may (in the long run) have unintended consequences for the sustainability of human livelihoods.

Land values reflect not only their agricultural productivity but also their value relative to other uses, such as housing, industry, or environmental amenities. Up until very recently, land values in the Imperial Valley—because the Valley was not gaining in population or experiencing economic growth—were relatively low. In contrast, in most other agricultural regions in California growers have been subject to pressure due to values capitalized on past profitability and to the logic of faster crop turnover, the other force affecting agricultural production. This later phenomenon is described as "appropriationism." Goodman et al. (1987) explain appropriationism as the process by which industrial countries have sought to reduce risk by modifying the natural reproduction of plants and animals, including developing tomatoes that will all ripen at the same time in a field or perfecting the factory farming of pigs and chickens).

In California, Guthman (2004) explains how farmers have changed their practices toward more careful pest management and continual cash cropping—through greater intensification of land use—and a transition in many places to higher-value crops, with far more sensitivity to market demand for products grown more "sustainably." This includes, as well, more sophisticated irrigation practices and techniques. The Imperial Valley, characterized by individual and corporate absentee landowners, inexpensive water, large acreages, and cheap labor, has maintained an agricultural system that today is increasingly under pressure due to recent combined changes in the Valley—higher land prices and urbanization, political power exerted by urban areas desirous of the agricultural water, pressure to reform wasteful water practices, and pressure from the environmental community concerned about the preservation and restoration of ecological processes and systems.

Imperial Valley agriculture seem to be suffering a kind of time lag, given that it has not been as subject to urban pressures and increasing land prices as compared to other agricultural regions in the state. As a 2004 Giannini Foundation report points out, California agriculture has been adept at meeting changing product demands, producing 350 crops in 2000, up from 200 crops in 1970. Other California agriculture has emphasized catering to affluent customers and reduced annual field crops in favor of higher-valued vegetable and perennial crops—nuts, fruits, ornamentals, nursery crops, and grapes. These crops require longer production cycles and significant planning and increase the state's vulnerability to market gluts and falling prices. Imperial Valley agriculture, by contrast, has continued to focus on more traditional crops, such as hay and cattle and vegetables. It has been able to do so, in part, because farmland prices have remained low, protecting growers from competition (*http://www.ccbres.sdsu.edu/data/indicators_of_the_month/pdf/2004/article_december.pdf*).

In 2006 the Imperial Irrigation District started a process it is calling the "Efficiency Conservation Definite Plan" to create a roadmap toward "wet water savings." By 2008 the district must conserve and transfer nearly 303,000 acre-feet of water annually—or nearly 1 out of every 10 acre-feet delivered each year from the Colorado River due to the QSA agreement (*http://www.definiteplan. com/intro-to-definite-plan.php*). To do so, the district must rely on voluntary actions—and only recently has it gone out to growers to solicit ideas. Curiously, therefore, the district seems to starting from zero—despite having a long-standing University of California agricultural research station located in its midst as well as the benefit of nearly a century of assistance on agriculturally related matters. The culture of the district, the relative autonomy of the agency and the isolation of the Valley, have seemingly insulated it from the modernization trends evident elsewhere. This is reinforced by long-standing absentee owners that have not been deeply involved in IID affairs or committed to modernizing agriculture in the Valley (Rudy, 1995). The cultural rules and norms, and the historical institutional arrangements of ownership concentration and absenteeism, were, we argue, not organized to act proactively in regard to the impending water regime changes brought about by growth in other basin states.

Still, the sustainability question remains. The Valley presents exceptional growing conditions in the market shed of very large cities—Los

Angeles and San Diego, as well as Phoenix, Arizona, and Mexicali on the Mexican side—but could the potential loss of agricultural lands represent a long-term sustainability loss relative to food production? What might that mean to the long-term ecological footprint of these cities as they come to have less potential of direct food importation from the Imperial Valley? The valley boasts two relatively rare features: its year-round growing climate with very high productivity and its proximity to huge urban markets in coastal southern California. If the price of oil continues to increase, the cost of food imported long distances is likely to increase as well. Though labor costs are much less in Latin America and Mexico, the combined efficiencies in California agriculture and shorter distribution distances make California agriculture competitive today—but what will happen in the future? Even if the region's cities became more dependent on more locally grown food, farming practices are structured and framed by an institutional regime in which production practices are highly formulaic—and tend to be high in fossil fuel and chemical inputs. The change in production emphasis and methods in other parts of California shows this is possible, but will absentee farm owners invest in those changes or find it more profitable to sell their lands for urban expansion? And if the changes are predicated on higher prices for farm products, then for whom will the food be produced? Moreover, the eventual effects of climate change on the flow of the Colorado River remain unknown, but a growing consensus points to a decline. Could more water efficiency on the farm compensate for this drop? Sustainability thinking requires engagement with uncertainty at multiple scales.

Decision Making in San Diego

Over a 75-year period, the water transfer from the Imperial Irrigation District to San Diego will provide 200,000 additional acre-feet of water per year. This amount of water is sufficient to meet the projected urban growth of approximately 1 million people by 2030 at the current water consumption rate of over 200 gallons per capita per day for all urban uses, and up to 280 per day in the summer. The price of water will start at $258 per acre-foot and will increase to $420 by year 15. The San Diego County Water Authority is a water retailer that delivers water to its member agencies, including cities, irrigation districts, municipal water districts, county water districts, and the Pendleton Military Reservation. As previously mentioned, currently, 70–95% of the county's water supply is imported from the Colorado River and northern California and the SDCWA has relied on the Metropolitan Water District of Southern California for imported water supplies. The total water delivered via the SDCWA system is 720,700 acre-feet. The transfer will also provide the MWD up to 110,000 acre-feet per year, and other transfers from agriculture to urban water use are also included in this highly complex arrangement, largely among water wholesalers.

The water transfer was based on projections of population growth in the region and was negotiated by the SDCWA. While the SDCWA sets water rates, direct water consumption management is not part of its administrative authority; it simply wholesales water—and the more it sells, the greater its profitability. It is solely an advocate for the implementation of best-management practices to obtain water conservation savings. Thus, the water transfer from the IID to the SDCWA did not entail any mandatory measures for urban water conservation. Moreover, the SDCWA 2005 Urban Water Management Plan projects a 12% water conservation rate by 2030 (*http://www.sdcwa.org/manage/pdf/2005UWMP/Final 2005UWMP.pdf*, pp. 2–5).

Yet, estimates of the potential to improve water use efficiency and conservation in urban water use in California show much greater potential. The Pacific Institute, a nonprofit think-tank, conducted a peer-reviewed study of the potential for urban water conservation in the state and estimated a potential of 33% greater savings. According to the Pacific Institute, the residential sector is the largest urban water user and offers the largest volume of potential savings. With current technologies and policies, residential water use could be as low as 60–65 gallons per day (gpd) without any change in the services actually provided by the water. In San Diego County, residential use ranges around 200 gpd. Even without improvements in technology, the Pacific Institute estimates that indoor residential water use could be reduced by nearly 40% by replacing remaining inefficient toilets, washing machines, showerheads, and dishwashers, and by reducing the level of leaks. With improved management practices and available irrigation technology, another 34% water savings could be effectuated on outdoor residential water use. They also found water savings could be realized in industry, such as paper and pulp, commercial laundries, schools and oil refineries, adding up to about a 39% potential (Pacific Institute, 2003). The SDCWA may argue that water pricing will eventually reduce use, as the price of outdoor irrigation may become prohibitive; yet, it appears that no real consideration of what an investment in conservation, equivalent to the ultimate additional costs of the transferred water (at $420 per acre-foot), might have yielded in newly available water. Between ½ and 1 acre-foot is estimated to be about what a family uses per year. However, for water agencies to institute water conservation poses a structural problem for their revenue stream—the less water that is sold, the less income they receive. Thus, a water conservation strategy can be seen as one that undercuts revenue. This is due to the way in which water districts were organized at the turn of the 20th century. The rules encouraged growth.

San Diego County now has about 3 million people, and 62% of the homes in the region were built after 1970. Under current plans and policies more than 90% of the vacant land designated for housing for the additional growth of 37% in the next 20 years is planned for densities of less than one home per acre (many of these are on wells); and of the remaining vacant land planned for housing in the 18 incorporated cities, only about 7% is planned for multifamily housing (SANDAG, 2004). Multifamily homes use less water and less land per unit. Much of the growth since 1990 has occurred in the northern part of the county, areas with high ecosystem diversity, including the endangered coastal sage scrub ecosystem. Land-use decision makers in the San Diego area are able to ignore water scarcity, as it is the responsibility of the SDCWA to provide water and because there are no substantive policy linkages. The regulatory structure is fragmented and siloed. Water districts are responsible for water, and cities and counties are responsible for their own land use planning—each to its own.

Coastal ecosystem environmentalist activists, trying to preserve the endangered coastal sage scrub, and those trying to preserve the Salton Sea have not been active in the water conservation arena. Open space advocates in San Diego County that have worked to curb sprawl by designating open space lands have not addressed the growth facilitating water transfer nor promoted water conservation (Endangered Habitats League newsletters, *http://www.ehleague.org*). They are focused on land-use decisions, and the siloed regulatory structure means they don't make the link with water policy.

The water transfer from the Imperial Valley—a nearby region—to San Diego interweaves the two areas' future, though the regions are treated as distinct and unrelated. Thus, it is quite plausible that the transfer will have sustainability effects in both places that involve negative impacts on nature and resources, as well as community development (most of the planned future growth in San Diego County is in the affluent rural areas at low densities), that would not have occurred had there been no water transfer or less water transferred. If there had been no additional water available in San Diego, development patterns would have had to take water availability into account, leading to potentially denser land-use development, more multiple-family developments, and greater water conservation technologies. The ancillary benefits would have been a more compact pattern of urban development, espoused by those working on urban sustainability strategies.

At the same time, for the Valley, no water transfer would have meant a continuation of less intensive and efficient agriculture, inefficient water use, and continued flows to the Salton Sea rather than fallowing.

The Colorado River Compact and the development of dams and irrigation systems on the river were in many ways the result of adept political lobbying by Imperial Valley residents, politicians, and the Imperial Valley Irrigation District in the early 20th century. Subsequently, due to the availability of Colorado River water, urban development was possible in Arizona and Nevada. Rules and conventions were created during the first half of the last century that favored the Imperial Valley and large-scale agriculture—but also suburbanization, with the Federal Highway Act, the creation of the Federal Housing Administration, and other federal programs (Rome, 2001; Fishman, 1985; Jackson, 1987). Policymakers probably did not anticipate an era in which these two forces might find themselves at odds.

Institutions, Spatial Scale, and Interregional Relations

Much of the water in the West was developed by the federal government in the first half of the 20th century and is regulated by federal policy. Surprisingly, perhaps, federal water in California is delivered by local water districts, a result of the late-19th-century decision by the state to devolve water development and delivery to local entities (Pisani, 1992; Pincetl, 1999). There is little direct linkage between federal water policy and state water policies and regulations. Changes in water deliveries to the IID were driven largely by urban growth in other basin states, reducing the surplus California had learned to depend upon and triggering the long-standing compact regulating interstate water allocations. At the same time, the rise in neoliberal approaches to resource management over 25 years shifted discourses about government-developed water to its transferability and marketability. The 1904 Reclamation Act has been reinterpreted to suit new cultural norms and values developed to support large-scale farming but also large-scale real estate development and most recently water transfers from regions, based on the ability to pay.

The Quantification Settlement Agreement in 2002 opened the door to selling water by water districts that, up until then, had previously been appropriated water for use in that district. Local districts can now operate as water entrepreneurs, selling their federally developed water for district revenue. Environmental organizations that had early on lobbied for water markets, thinking it would create more water for the environment, in this situation facilitated the consolidation of power and the income stream of the IID. Environmentalists conditioned the water transfer to include "environmental" water for the Salton Sea at the expense of borderland wetlands in Mexico (with the planned lining of the All-American Canal) and lack of concern for the Colorado River delta. Additionally, they did not take into consideration the effect of urban growth in San Diego that will be enabled by more water on endangered habitats in that coastal region. There was no involvement of watershed groups, or other water-oriented quasi-governmental organizations in the Imperial Valley and none in San Diego either. In part, we believe, this is due to the isolation of the Imperial Valley from the rest of California and the lack of compelling natural scenery (or endangered species), other than in the Salton Sea.

Existing structures of political decision making facilitated the process as it evolved. First there is the fragmentation of decision making in the area of water, where local water districts are autonomous in their processes from the counties and cities they serve and from state government. Second, cities and counties themselves need not coordinate decision making among themselves, and federal decision making is also entirely autonomous.

While this is a classic problem in a federalized political system whose organizing principle is that of decentralized control and fragmentation of authority, potential problems were compounded in this case by few, if any, formal institutional bridges, such as interagency councils. All agencies are out for their own interests. The Colorado River Board, created in 1937 by state statute to protect riverine states rights and interests in the resources provided by the Colorado River and to represent states in discussions and negotiations regarding the Colorado River and its management, has representatives from seven counties in southern California. The state itself is represented only by an ex officio member, and there is no juridical/legal leverage that the board has relative to decisions of the federal government; it is advisory only. Within the regions them-

292 THE SUSTAINABLE DEVELOPMENT PARADOX

selves, water authorities and districts depend on water sales for revenues, and therefore relations are competitive. They are in a zero-sum game situation. If aggressive conservation takes place, they will sell less water and reduce their fiscal viability. If one district reduces its water use, another district will receive more water and make more sales. County government water retailers have no linkages with the water districts, but each county and city is in competition with the other for growth. Though the state legislature now requires all cities and counties to develop an urban water management plan (updated every 5 years), these have no regulatory authority.

Thus, we have a regulatory framework that has inherent constraints propelled by specific institutional needs (selling more water, for example), and intense competition, where one's loss is another's gain. The regulatory regime has been shaped and reinforced over time by powerful interests that derive profit and long-term viability from the ways in which water and land-use policy is made. Going into the future, however, there are several factors that are now less predictable; among them, the effects of climate change on Colorado River flows and the availability and price of oil. The amount of water that flows in the river—while it can be regulated and stored—ultimately is a material factor that will impact the future mode of production and the regulatory regime that has been built up over the past century. How climate change will affect rainfall in the states through which the river flows is still unknown, though most predictions are for drier conditions. This will further reduce the flow of the Colorado River and intensify competition over its allocation, both among states and between agriculture and urban uses. Oil supply and price fluctuations could send shocks through modes of production and regulatory regimes as well.

In the Imperial Valley there is an agricultural sector that successfully defeated the United Farm Workers Union and other social movements into the early 1980s. But by the time the U.S. Department of the Interior came to the table to get California to reduce its surplus use of the Colorado River by the late 1990s, joined by the Metropolitan Water District, the San Diego County Water Authority, and the governor of California, the district and the valley did not step forward to defend agriculture. The IID stood to make considerable money on the transfer, as the transferred water was to be compensated for at historically record prices.

The county government and the cities will not share any of the financial rewards provided to the IID for fallowing and future water conservation measures, and nor do they have any particular vested interest in the maintenance of agriculture in the Imperial Valley in the face of the potential for land development and increasing property values, even though the county went on record as opposing the water transfer (*http:// www.imperialcounty.net/supervisors-agendas/2004/2004% 20Resolutions/ 2004-34.pdf*). In any event, both the cities and the county were left out of the deal making by the IID. As Stephen Elkin (1987) points out, city politics is a profoundly economically oriented enterprise, and promoting economic growth in the city has come to mean the city as a pattern of land use. For Imperial County and its cities to prosper, they benefit most from a transition to urbanization, given the current structure of agriculture and the taxation structure of state and local governments in California.

Yet, the water transfer has left the valley bitterly divided, and the 2006 IID board election replaced protransfer representatives with ones who are opposed to the transfer. Residents and decision makers now feel like victims, as though the water transfer was done *to* them rather than *by* them. Resident farmers, including some of the largest farmers, organized to form the Imperial Group. The group claims that the farmers are the real owners of the water and that the IID has outlived its usefulness as an irrigation district (see *http://www.imperialgroup.info/ info_9pt.php*). The Imperial Group has sued the IID, challenging the right of the IID to enter into an agreement regarding transferring water without participation by the landowners, since the Imperial Group claims the landowners hold the water rights.

CONCLUSION

The case of the Colorado River water transfer to San Diego County points to the need for greater investigation of the domains and scales with which sustainability must contend. As Sneddon suggested, sustainability can only be considered relative to given socioeconomic contexts and specific geographical locations. We try to show the structures, norms, and values that have defined the rights, constraints, and power that influenced the development of the regions and that would need to be addressed to shift toward greater sustainability—yet to be defined. The regulatory framework that exists is a historically accreted tangled complexity that operates on several interlinked geographical scales. It is defined by multiple distinct institutions that regulate at different jurisdictional scales in quasi-autonomous self-reproducing spheres. At the heart of this complexity lies water and land, the material substrate utilized to produce human sustenance. We observed how diffuse, conflicting, and decentered "authority and power are inscribed across the intermingling domains of nature and culture" (Whiteside, 2002: 114).

Selling federally developed and subsidized water from the Imperial Valley—where it was historically designated for agriculture but eventually redirected at much higher prices to an adjacent urban area—represents a significant sea change in the institutions of water management through the application of private sector thinking. Bakker (2002) terms this the commercialization or mercantilization of water. It is premised on a new universal condition of water scarcity, simultaneously natural—justifying cost-reflective pricing—and social, the result of flawed public management (in our case inefficient water management by the IID). As Bakker notes (2002: 772), this is a process initiated not by the market but by the state and thus should be understood as a process of reregulation rather than deregulation, that is, characterized by an emergent new form of governance of natural resource allocation. We have tried to show how this process of reregulation of the social metabolism of nature that is taking place is highly contingent on local variables, including the manmade Salton Sea and its environmental advocates, a highly concentrated corporate and absentee landlord form of agriculture, and a fragmented governmental structure with little or no coordination or collaboration. The environmental movement in California and its concern about water use by agriculture also contributed to the legitimacy of transferring water from agricultural to urban uses (Pincetl, 1999).

Our political ecological approach emphasized the location-specific biophysical materiality of water and land and how they have been appropriated and configured to suit the mode of production, including ownership rules. Using this approach and fusing a regulationist analysis to understand today's trend toward mercantilization of water resources, we suggest, is the starting point for a sustainability analysis. While we have only sketched out in barest detail the complex and intertwined characteristics of land and water in Imperial County in California, we hope to have made the case that for a more sustainable regulatory regime of land and water use across scales, understanding the organizational and institutional specificities over time and across space is indispensable. We show how the introduction of new allocation rules—in our case, a water transfer—has engendered a new hydrosocial contract between users and their environment and a new exchange value to the resource. An additional economic analysis based solely on ecological economics would likely argue against the long-term sustainability of this water transfer relative to truly efficient water use, and the net negative implications for both the environment and urbanization trends.

The new water transfer contract also has long-term, and yet uncertain, implications for agriculture in the Imperial Valley as well as for urbanization in both the Imperial Valley and San Diego County. It introduces market-simulating decision-making techniques, shifting water from its status as a public resource for public benefit to that of a liquid commodity that now embodies new values, such as highest and best use and efficiency, reflecting a neoliberal turn of events. The environmental and human dimensions of this change and its ultimate sustainability are unlikely to lead to favorable outcomes.

ACKNOWLEDGMENTS

We wish to thank Michael Storper, Andy Jonas, Donald Worster, Donald Pisani, and David Gibbs for their thoughtful comments and suggestions.

NOTE

1. Both of us have spent time in Imperial County. Stephanie Pincetl is a party to a lawsuit against the lining of the All-American Canal through her membership on the board of a nonprofit organization, Citizens United for Resources and the Environment (CURE). The reconstruction of this timeline and potential impacts comes from her immersion in the county over a 3-year period. Neither of us is financially involved with CURE.

REFERENCES

Bakker, K. (2002). From state to market? water *mercantilización* in Spain. *Environment and Planning A, 34*, 767–790.

Barclay, B., Schmidt, J., & Hill, D. (1980). State, capital and legitimation crisis: Land and water in California's Imperial Valley. *Contemporary Crises, 4*, 1–26.

Beatley, T. (1995). Planning and sustainability: The elements of a new (improved?) paradigm. *Journal of Planning Literature, 9*, 383–395.

California Agriculture. (1997, May–June). Challenge, promise for nation's "winter salad bowl," pp. 4–5.

Carle, D. (2004). *Introduction to water in California*. Berkeley: University of California Press.

Cleveland, C. (1987). Biophysical economics: Historical perspective and current research trends. *Ecological Modeling, 38*, 47–73.

Cleveland, C. (1991). Natural resource scarcity and economic growth revisited: Economic and biophysical perspectives. In R. Costanza (Ed.), *Ecological economics: The science and management of sustainability* (pp. 289–318). New York: Columbia University Press.

Cleveland, C., Stern, D., & Costanza, R. (2001). *The economics of nature and the nature of economics*. New York: Edward Elgar Publishing.

Coates, P. (1998). *Nature, Western attitudes since ancient times*. Berkeley: University of California Press.

Cory, H. T. (1915). *Imperial Valley and the Salton Sink*. San Francisco: J. J. Newbegin.

Costanza, R., Cumberland, J., Daly, H., Goodland, R., & Norgaard, R. (1997). *An introduction to ecological economics*. Boca Raton, FL: CRC Press.

Costanza, R., & Daly, H. E. (1992). Natural capital and sustainable development. *Conservation Biology, 6*, 37–46.

Costanza, R., Farber, S. C., & Maxwell, J. (1989). Valuation and management of wetland ecosystems. *Ecological Economics, 1*, 335–361.

Drummond, I., & Marsden, T. (1999). *The condition of sustainability*. London: Routledge.

Elkin, S. (1987). *City and regime in the American republic*. Chicago: University of Chicago Press.

Escobar, A. (1995). Imagining a post-development era. In J. Crush (Ed.), *Power of development* (pp. 211–227). London: Routledge.

Fine, H. (2002, July 1–7). LA's search for water, a special report. *Los Angeles Business Journal*, pp. 1, 10, 11.

Fishman, R. (1985). *Bourgeois utopias, the rise and fall of suburbia*. New York: Basic Books.

Glacken, C. (1967). *Traces on the Rhodian shore*. Berkeley: University of California Press.

Goldschmidt, W. R. (1946). *Small business and the community: A study in the Central Valley of California of effects of scale of farm operations*. Report of the Smaller War Plants Corporation to the Special Committee to Study Problems of American Small Business. Washington, DC: U.S. Government Printing Office.

Goodman, D., Sorj, B., & Wilkinson, J. (1987). *From farming to biotechnology: A theory of agro-industrial development*. New York: Blackwell.

Gottlieb, R., & FitzSimmons, M. (1991). *Thirst for growth: Water agencies and hidden government in California*. Tucson: University of Arizona Press.

Guthman, J. (2004). *Agrarian dreams, the paradox of organic farming in California*. Berkeley: University of California Press.

Hundley, N. (1992). *The great thirst: Californians and water, 1770s–1990s*. Berkeley: University of California Press.

Imperial Irrigation District. (2004). Resolution 2-2004 of the Board of Directors of the Imperial Irrigation District (IID) authorizing execution of the agreement of purchase and sale dated February 2004, agreement of purchase between Western Farms L.P., a California limited partnership, (as seller) and Imperial Irrigation District (IIS) as purchaser. Accessed April 24, 2005, at *www.iid. com/water/ fallowing_program.html*.

Jackson, K. T. (1987). *Crabgrass frontier, the suburbanization of the United States*. New York: Oxford University Press.

Klug, H. (2002). Straining the law: Conflicting legal premises and the governance of aquatic resources, *Society and Natural Resources, 15*, 693–707.

Lélé, S. (1991). Sustainable development: A critical review. *World Development, 19*(6), 607–621.

McWilliams, C. (1949). *California the great exception.* New York: Current Books.

McWilliams, C. (1968). *North from Mexico: the Spanish-speaking people of the United States.* New York: Greenwood Press. (Originally published in 1948)

Pacific Institute. (2003). *Waste not, want not: The potential for urban water conservation in California.* Accessed April 23, 2005, at *www.pacinst.org/reports/urban_usage/.*

Peet, R., & Watts, M. (1996). Liberation ecology: Development, sustainability and environment in an age of market triumphalism. In R. Peet & M. Watts (Eds.), *Liberation ecologies: Environment, development, and social movements* (pp. 1–45). London: Routledge.

Perkins H., & Thorns, D. (2001). A decade on reflections on the Resource Management Act 1991 and the practice of urban planning in New Zealand. *Environment and Planning B: Planning and Design, 21*(5), 639–654.

Pincetl, S. (1999). *Transforming California: A political history of land use and development.* Baltimore: Johns Hopkins University Press.

Pisani, D. (1992). *To reclaim a divided west: Water, law and public policy, 1848–1902.* Albuquerque: University of New Mexico Press.

Redclift, M. (1987). *Sustainable development: Exploring the contradictions.* London: Methuen.

Reisner, M. (1986). *Cadillac desert.* New York: Penguin Books.

Revised Fourth Amendment agreement between Imperial Irrigation District and San Diego County Water Authority for transfer of conserved water, October 2003. (2003). Accessed April 24, 2005, at *http://www.sdcwa.org/manage/mwdQSAdocs.phtml.*

Robinson, J. (2004). Squaring the Circle?: Some thoughts on the idea of sustainable development. *Ecological Economics, 48*, 369–384.

Rome, A. (2001). *The bulldozer and the countryside.* Cambridge, UK: Cambridge University Press.

Rudy, A. P. (1995). *Environmental conditions, negotiations and crises: The political economy of agriculture in the Imperial Valley of California, 1850–1993.* PhD dissertation in sociology, University of California, Santa Cruz.

San Diego County Water Authority. (2004, July). Regional comprehensive plan for the San Diego region, urban water management plan. Accessed November 9, 2006, at *http://www.sdcwa.org/manage/pdf/2005UWMP/Final2005UWMP.pdf.*

Sayer, A. (2000). *Realism and social science.* London: Sage.

Sneddon, C. S. (2000). "Sustainability" in ecological economics, ecology and livelihoods: A review. *Progress in Human Geography, 24*(4), 521–549.

State of California, Department of Water Resources. (2004). Colorado River QSA & SS ecosystem restoration legislation. Accessed March 4, 2004, at *http://www.saltonsea.water.ca.gov/legis.indes.cfm.*

State Water Resources Control Board. (1984). Imperial Irrigation District Alleged Waste and Unreasonable Use of Water Order : 84–12 Affirming Decision 1600 and Denying Petition for Reconsideration. Available at *http://www.waterrights.ca.gov/hearings/water rights.*

Tempest, R. (2002, October 19). Proposed water sale makes division in desert. *Los Angeles Times*, p. A 1.

Veolia Environnement. (2006). U.S. Filter unit of Veolia Environnement announces sale of farmland in California. Accessed August 7, 2006, at *http://www.veolia water.com/access/press/?news=699&year=2004&geo=2*.

Waller, T. (2004). Expertise, elites, and resource management reform: Resisting agricultural water conservation in California's Imperial Valley. *Journal of Political Ecology, 1*, 13–41.

Whiteside, K. H. (2002). *Divided natures, French contributions to political ecology.* Cambridge, MA: MIT Press.

Williams, R. (1980). Ideas of nature. In R. Williams, *Problems in materialism and culture* (pp. 67–88). London: Verso Editions and New Left Bookclub.

World Commission on Environment and Development. (1987). *Our common future.* Oxford, UK: Oxford University Press.

Worster, D. (1985). *Rivers of empire, water aridity and the growth of the American West.* New York: Pantheon Books.

Worster, D. (1993). *The wealth of nature, environmental history and the ecological imagination.* Oxford, UK: Oxford University Press.

Yniguez, R. (2004a). Collins to keep entity on track. Accessed July 29, 2004, at *http://www.IVPressonline,IVPressonline.com*.

Yniguez, R. (2004b). Farmers' group fears fallowing could lead to $190 million loss. Accessed November 12, 2004, at *http://www.IVPressonline*.

Zimmerman, J. (2001). The "nature" of urbanism on the new urbanist frontier: Sustainable development, or defense of the suburban dream? *Urban Geography, 22*(3), 249–267.

Index

Aalborg Charter, 136
Actor Network Theory, 254, 260n7, n9
Affordable housing, Massachusetts statute, 113
Agriculture (Imperial Valley, California)
    employment, 284
    future of, 285
    land use relative value, 284
    rural–urban relations, 273–275
    transformation of, 271
    unique characteristics, 284–287
    water transfer effects, 273–280
Airport expansion, Barcelona, 140–141, 154n10
All American canal, 282–283, 295n1
Apocalypse, media headlines, 16
Atlanta, Georgia case study, 148–149
Austin, Texas, 104–109, 114–116
    governance, 114–116
    greening of, 104–109
    high-tech boom, 104
    as new economy space, 97–98, 104–109,
        114–116
    problems in, 105–107
    quality-of-life, 105–107
    Smart Growth Initiative, 106
Austin Green Builder Program, 104–105
Austin Network, 107–108

"Back Bay," Boston, 114
Barcelona, 131–142, 160–188; See also
        Metropolitan Region of Barcelona
    airport expansion, 141–142, 154n10
    entrepreneurial city case example, 131–142
    image of, 137
    integrated sustainability, 160–188
    propaganda events, 136, 153n7
    public spaces, 160–188
    social struggle, 138–141
    sustainability politics, 124, 167–170
    tensions, 138–142
    urban sustainability problems, 131
    urban transformation, 132–134, 153n5
        sustainability role in, 134–138
Barton Creek, Austin, Texas, 114

Best Value program, 196, 201–202
Biological diversity
    and fragmentation model, 246
    inclusive definition, 238, 259n1
Birmingham, England case study, 252–253, 255
Boston, Massachusetts, 109–116
    formula for success, 99, 109–113
    governance, 114–116
    knowledge-based economy, 109
    "Metro-Future" project, 112–113
        as new economy space, 97–98, 109–113
    "smart growth," 114–115
"Boston Formula," 111, 114
Boston 400 plan, 110–111
Bourn Brook River, 252–253
Bruntland Report
    American "smart growth" response, 206
    community involvement emphasis, 192
    key concepts, 45
    mantra of, 98
    neoliberal capitalism response to, 44–46
    "sustainable development" in, 43, 259n2
    urban governance principles, 128

California. See Imperial Valley, California
Cambridge, England, politics, 147
Capitalism. See also Neoliberalism
    ecological modernization as solution, 41–
        62
    limits to growth problem, 43–49
    Millennium report, 49–52
    neoliberal approach, critique, 41–49
    and new economy spaces, sustainability,
        117–118
    and spatial metaphors, 258
Carbon emissions, urban reduction strategies,
        124, 153n1
Case study arena, 3–4
Central Arizona Project, 277, 279–280
Central government. See also Governance
    community initiatives link, 233
    economic growth emphasis, 143, 154n15
    and Karvia, Finland forestry plan, 251

Central government (*continued*)
    sustainable communities role, 229–232
    in water development, California, 273–275,
        290–293
Central Valley Project, 274
Children, toxin vulnerability, 80–81
Circulatory tropes, 258
Cities. *See* Urban *entries*
Cities for Climate Protection, 136
Citizens United for Resources and the
        Environment (CURE), 282, 295n1
Citizenship
    public spaces function, 184–186
    sustainable communities link, 221
City planning. *See* Urban planning "Le Ciutat
        Sostenible"
    Barcelona, 132–142
    exhibit title, 136
    major moments in, 134
    tensions, 138–142
Climate change, politics, 21
Clos, Joan, 132, 137, 138, 140, 141
Club of Rome report, 43–46
Coachella canal, lining of, 283
Coachella Valley Water Authority (CVWA),
        280–281
Collective consumption
    in entrepreneurial city regions, 147
    in sustainable cities, 127–128
Collective provision, politics, 146–147
Collective space
    in high-density cities, 177–178
    in low-density model, 177
Colorado River
    and climate change factor, 272
    Imperial Valley water savings plan, 286
    Imperial Valley water transfer, 266–295
    salinity threat, 276
    San Diego water transfer, 287–290
    water allocations, 276–277
    water diversion effect on delta, 282, 291
    water transfer decision-making, 278–290
Colorado River Board, 291
Colorado River Compact, 290
Community development. *See* Sustainable
        communities
Community involvement
    British efforts, 193, 231
        countertrends, 231
    Bruntland Report emphasis, 192
    countertrends, 231
    nation-states link to, 233
    political modernization effect on, 196–199
    South East England case study, 200–201
Community 2020, 210
Compulsory competitive tendering, 195–196
Consensus
    key feature, 26
    in postdemocracy, 29
Conservation planning, 238–261
    components, 241–242
    "conviviality," 259
    European Union, 249–251
    fragmentation model, 241–259
        political power link, 256

function, 239–240
    increasing interest in, 239, 260n3
    Karvia, Finland case study, 250–251
    models of, 239–240, 246–249
    nonequilibrium model, 248
    regionalism, 249
    scale factor, 257–258
    spatial politics, 238–261
    Vincent Drive, England, case study, 252–253,
        255
"Contraction and convergence," 89
Corporate elites, sustainability response, 49
Corridors. *See* Wildlife corridors
"Cross-domain" analysis, 4–5
Cyborg urbanization, 59–60, 62n2

Democracy. *See also* Postdemocracy
    essence of, 35
    local government influence requirement, 193
    versus populism, 35
    and public spaces, 185
    in sustainable communities agenda, 230
Development, versus growth, definition, 48,
        62n1
"Dialogic democracy," 31
Discourse
    versus implementation, 225–226, 232
    and sustainable communities, 219–221, 225–
        226, 232
Drip irrigation, 283

Ecological economists
    extended balance sheet, 271
    water transfer application, 294–295
Ecological footprints
    implications, 4
    and local sustainability, 3–4
    in South East England policies, 206
    weaknesses, 57–58
Ecological modernization, 41–62, 201–202. *See
        also* Neoliberalism
    critique of, 41–43, 46–49
    environmental justice antipode, 60–61
    and New Labor politics, 201–202, 208
    "smart growth" policy, 207
    and survival of capitalism, 46–49
    in United Kingdom, 201–202
Ecological succession model, 245–246
Ecology. *See* Conservation planning; Landscape
        ecology
Economic development. *See* Capitalism;
        Neoliberalism
Economic disadvantage, feminization of, 76–82
Economics, and sustainability models, 2–3
Edge habitat
    connectivity, 241
    landscape ecology function, 241–242
Education gap, gender inequalities, 78
Employment, gender inequalities, 78
England. *See also* South East England
    Local Agenda 21, 202, 209
    local government restructuring, 194–199,
        209
    "political modernization," 192–210
    selective "sustainability fix" method, 153n3

sustainable communities reassessment, 214–237
sustainable development weakening, 192–213
English Heritage, 228
English Nature, 227
Entrepreneurial city, 123–159; See also New economy spaces
  Barcelona case example, 131–142
  boundary blurring, 130
  capitalist model differences, 127
  characterization, 125–127
  collective provision politics, 146–17
  cross-class alliances, 147
  governance, 125–131
  greening of, 123–159
  mobility politics, 147–149
  social issues, 144–145
  socially just food policy, Canada, 149–150
  spatial politics, 145–150
  sustainability challenges, theory, 143, 154n13, n14
  sustainability politics, 123–159
  tensions, 129–131
  theoretical narratives, 126
Environmental Defense Fund, 279–280
Environmental groups
  gender inequality, 75–76
  South East England case study, 200–201
  sustainable communities objections, 227–228
Environmental justice
  gender issues, 67–68, 76–82
  in radical urban political ecology, 60–61
  in sustainable development theory, 98–99
Environmental modernization. See also Ecological modernization
Environmental populism. See Populism
Environmentalism. See Conservation planning
Equitable sustainability, 4–5, 98–103. See also Social justice
European Commission
  socioeconomic inequalities, 221, 234n2
  sustainable development as fundamental objective, 220, 234n1
European Union
  conservation planning, 249–251
  gender inequality, 72
  gender mainstreaming, 82–83
    waste management, 83–88
  South East England development objections, 228
European Water Framework Directive, 249

"Factor Four," 210
Fallowing
  effects of, 282–283
  Imperial County, California, 282
Federal water development
  California, 273–275
  Imperial Valley, California, 290–293
FIRE sector, 95
Food policy, entrepreneurial city, 149–150
Fordism
  change to post-Fordism, 170
  versus the new economy, 100–101
Forestry, Karvia, Finland case study, 250–251

Fragmentation model, 241–259
  in conservation planning, 241–246
  ecological critique, 243–246
  ecological research influence on, 245–246
  effects of, 256–257
  and habitats, 241–242
  implementation, 253–254
  versus "incomplete boundaries," 247
  institutional power link, 256–257
  Karvia, Finland case study, 250–251
  linear connectivity logic, 258
  new conception of, 246–249
  persistence of, 248
  in practice, 249–253
  regulatory control amenability, 254–258
  spatial control facilitation, 254–256
  spatial politics of, 249–257
  "storyline" aspects, 256
  topological character of, 254–255, 260n9
  Vincent Drive, England case study, 252–253
Front Marítim in Barcelona, 133, 135

Gardened public space
  in low-density versus high-density cities, 176–178
  in Metropolitan Region of Barcelona, 177
  water consumption, 179–181
    versus private space, paradox, 179–181
Gender impact assessment, 82
Gender inequality, 66–94
  in business and industry, 74–75
  environmental negative effects, 67–76
  European Union policies, 82–83
  in government positions, 72–74
  and "microruptures," 67
  poverty issue, 76–78
  social disadvantages, 76–82
  in "sustainability" dynamic, 70–71
  toxin exposure vulnerability, 80–81
    pregnancy, 80–81
  and waste management case example, 83–88
  World Bank corrective policies, 82
Gender mainstreaming, 68–69
  definition, 82–83
  European Union policies, 82–83
  "microrupture" effect, 87–88
  obstacles, 87, 89–90
  in waste management, 82–88
Gentrification, in Barcelona region, 172
Gini Index scores, 221, 234n2
Global warming
  apocalyptic media headlines, 16
  Barcelona's contribution, 136
  and sustainable communities, 231
Globalization
  public spaces impact, 170–171
  "second modernity," 171
GLOBE ("go local on a better environment") groups, 200
Governance
  collective provision politics, 146–147
  community-based, 230
  entrepreneurial cities, 125–131
  in integrated sustainability, 167–170
  in new economy spaces, 114–116

Governance (*continued*)
    public spaces role, 184–186
    "second modernity" effect on, 171–172
    spatial policy, England, 226–228, 230–232
    and sustainable communities, 226–228
        new agenda, 230–232
Government Office for the South East (GOSE)
    central government's role in, 205
    mission of, 203, 207
Green practices
    gender inequities, 78–79
    social control function, 144–145
Greenhouse gas emissions, Barcelona, 136
Growth model, versus development, definition,
        48, 62*n*1
"Growth vortexes," 128

Habitat patches
    ecological critique, 244–246
    fragmentation concept, 241–246
    landscape ecology function, 241–242
    Vincent Drive, England case study, 252–253
Habitats Directive, 242, 249–250
Health issues, urban sustainability politics, 152,
        155*n*21
Heritage organizations, 228
House-building rates, South East England, 226
Household waste management, gendered
        aspects, 86–87
Hurricane Katrina, 145, 155*n*19
Hydrosocial contracts, 9, 294

Immigrants, Barcelona protests, 139–140
Imperial Group, 293
Imperial Irrigation District (IID)
    formation of, 275
    function, 275
    governance arrangement, 274–275
    interregional relations, 290–293
    profit motive effects, 291–293
    water transfer role, 279–284, 287
    "wet water savings" plan, 286
Imperial Valley, California, 266–295
    agricultural methods time lag, 286–287
    case study, 272–295
        introduction to, 275–278
    commercialization of water transfer, 294
    exploitation of land, 278
    farming production in, 277–278
    land development, 283
    land values, 285
    location, 275
    poverty in, 278
    regulatory bodies fragmentation, 290–292
    San Diego water transfer, 287–290
    social characteristics, 276
    water speculators, 283–284
    water transfer, 266–295
        decision-making, 278–292
Implementation, versus discourse, 225–226,
        232
Income, gender inequality, 77–78
Individualization
    low-density urban model, 175–178
    in "second modernity," 171–172
Industrial companies, gender inequality, 74–75

Institutional sustainability, 166. *See also*
        Political sustainability
"Insularity" of habitats, 241–242
Integrated Regional Framework, 204, 206
Integrated sustainability, 160–188
    Barcelona public spaces, 160–188
    economic dimension, 169
    environmental dimension, 168
    political dimension, 167–170
    in public spaces, 167–170
    social dimension, 165–170
Ireland, gender mainstreaming, 83–84
Irrigation programming, gardens, 180
Island biogeography, 247, 251

Just sustainability, 4–5, 98–103. *See also* Social
        justice

Karvia, Finland
    case study, 250–251
    fragmentation model in, 252–253, 255
Katrina Hurricane, 22
Key Worker Living Programme, 223
Keynesian policies, 222
Kyoto Protocol, postpolitical aspects, 30

Labor force, fragmentation in post-Fordism,
        170
Landscape ecology. *See also* Spatial politics
    components, 241–242
    European Union plan, 249–252
    fragmentation model, 241–257
        critique, 243–246
    new models, 246–249
    nonequilibrium models, 248
    political versus scientific realm, 248, 260*n*5
    urban planning, fragmentation model, 251–
        253
        Vincent Drive, England case study, 252–
        253
Local Agenda 21
    in Barcelona, 131, 136
    British central government lack of support
        for, 202, 209
    conceptual background, 3
    and England's New Labor government, 196,
        198, 202, 209
    South East England case study, 200, 209
    and Sustainable Community Strategies, 198
    urban acceptance, 129
Local governments
    capacity for action debate, 5–6
    Comprehensive Performance Assessment
        effect on, 199, 209
    decline of influence, England, 194–201, 209
    environmental strategy impediments, 193,
        209
    and Karvia, Finland forestry plan, 251
    New Labor government influence on, 196–
        201, 209
    political modernization effects on, 196–201
    sustainability initiatives, 1
    in sustainable communities agenda, 230
    "well-being" promotion, England, 198–199
    women in, 73–74
Local Strategic Partnerships, 198, 201

London First, 227
Low-density urban spaces
    ecological problems, 175–176
    garden water consumption, 179–181
    individualization in, 175–178
    political effects, 187–188
    public spaces usage, 182–186
    social effects, 175, 182, 186–188

Managerial approach, in postpolitics, 24
"Manchester of Catalunya," 137
Market produced values, 2–3
Marxism, political ecology link, 71
Massachusetts, new economy initiatives, 113
Media headlines, global warming, 16
Metabolism of cities. See Urban metabolism
"Metro-Future" project, 112–113, 115
Metropolitan Region of Barcelona, 160–186.
        See also Barcelona
    direct observations, 165
    economic integration, 168–169
    gardened public space, 176–177
    integrated sustainability, 160–188
    location maps, 163–164
    low-density spaces impact, 175–178
    "methodological triangulation," 163–164
    new urban model negative effects, 172–173
    political integration, 167–170
    public garden unsustainability, 179–181
    public spaces, 160–188
        recent transformation, 176–178
        political sustainability, 184–186
        social sustainability, 181–184
    qualitative work, 164–165
    quantitative work, 163–164
    scalar integration, 173–178
    "second modernity" impact on, 173–186
    social integration, 165–170
    transverse integration, 178–186
    urban lifestyle implications, 173–178
    urban sprawl effects, 172–176
Mexico, Colorado water transfer effects, 282–
        283
Microgeographies
    gender issues in, 67–68, 78
    in household waste management, 86
"Microruptures"
    gender equality link, 67–68, 87–88
    political ecology implications, 71
Millennium Ecosystem Assessment, 49–52
Mobility, and urban politics, 147–149
Models, of sustainability, 2–4
Modernization. See Ecological modernization;
        Political modernization
Multifamily homes, water consumption, 289–290
Municipal waste disposal. See Waste
        management

Natura 2000 network
    fragmentation model in, 254
    Karvia, Finland case example, 250–251
    purpose of, 242, 249–251
    Vincent Drive, England example, 252–253
Nature conservation. See Conservation planning
"Natures," 14–38
    indeterminacy, 17–20

versus neoliberal capitalist view, 6–7, 19
    question of, 14–17
Neoliberalism. See also Capitalism; Ecological
        modernization
    critique, 41–49
    versus democracy, 36
    and England's "Best Value" initiative, 202
    England's New Labor's link, 208
    in entrepreneurial cities, 126–128, 153n2
    as fiction, 36
    and government water transfers, 208
    inclusive politics as answer to, 31–32
    inevitability, 24
    and new economy spaces, 117–118
    postpolitical postdemocracy view of, 34–36
    postpolitics link, 24
    versus radical urban political ecology, 57–61
    "smart growth" link, 207
    versus sustainable development, England,
        194
    view of nature, 6–7, 19
"Network governance mode," 108
New economy
    definition, 100
    economic development in, 100–101
    versus Fordism, 100–101
    governance, 101–103
    growth pressures, 102
    local space embeddedness, 100–101
    problematic issues, 101–103, 116–117
New economy spaces. See also Entrepreneurial
        cities
    Austin, Texas, 104–109
    Boston, 97–98, 109–116
    contradictions in, 116–118
    development concept in, 100–101
    governance, 114–116
    neoliberalism influences, 117–118
    top 10 U.S. urban cities, 97
    in urban development, 95–122
New Labor government
    environmental modernization, 201–202, 208
    political modernization, 196–201
    South East England role of, 205
"New Localism," 3
New Orleans's Katrina, 22
New regionalism
    implementation challenge, 233
    net effect of, 223
    rise of, 221–223
    socioeconomic inequality result of, 223
    spatial policy impact, 222, 233
    and sustainable communities, England, 221–
        223
Nineteenth century environmental reforms, 151
Nonequilibrium ecology, 247–248
"Nonhuman" category, 254, 259, 260n7
Northern Arc debate, 148–149
Nutrition, gender inequality, 77–78

Opportunity Austin, 107–108
Organization for Economic Cooperation and
        Development, 129–130
The Pacific Institute, 288
Parc del Llobregat, 140
Parliamentary rule, in postpolitics, 25, 29

Participatory government
  and public spaces, 184–186
  urban sprawl effects on, 187
Patches. *See* Habitat patches
Patriarchy, and "sustainable development," 70
Pay equity, gender inequality, 77
Poble Nou, 133, 153n6
Political ecology
  fragmentation model facilitation of, 254–
    256
  social change implications, 71
  in urban environment, 247–249
    dynamics, 248–249, 260n6
  and water transfer, Imperial Valley, 294
"Political modernization," 196–201
  in Britain, 192, 210
  local government effects of, 196–199
  negative implications, 193
Political sustainability
  importance of, 167
  in integrated sustainability, 167–169
  public spaces in, 184–188
Politics. *See also* Postpolitics
  of climate change, 21
  of collective provision, 146–147
  entrepreneurial cities, 125–127, 143–144
  and environmental fragmentation model,
    254–256
  fundamental nature of, 25
  gender inequality, 72–74
  nature in, 21
  in new economy spaces, 95–118
  versus postpolitics, 24–25
  and public spaces, 161, 184–186
  in radical urban ecology, 55–61
  and science, need for integration, 54–55
Populism, 32–35
  versus democracy, 35
  environmental concerns, 32–35
  externalized and objectified enemy in, 33
  fundamental fantasy of, 33–34
  in postpolitical politics, 34
  tactics of, 33–35
Portugal, gender mainstreaming, 83–84, 90
Positivist science, 51–52
Postdemocracy, 27–32
  definition, 28
  "dialogic" mode in, 31
  populism in, 34
  postpolitics link, 27–32
Post-Fordism, 170
Post-humanism, 59, 62n3
Postpolitics, 13–40
  assumptions, 31–32
  configuration of, 30
  consensus in, 26
  constitution, 30
  definition, 23–25
  environmental politics as, 22–23
  versus genuine politics, 24–25, 30
  managerial approach in, 24
  neoliberal politics link, 24
  parliamentary rule in, 25
  populism in, 34
  postdemocracy link, 27–32
Poverty, feminization of, 76–78

Power relations
  environmental consequences, 37–38
  fragmentation model, 256–257
  gender inequality, 72–74
  Imperial Valley, California water transfer,
    266–295
  in "sustainable development," 70
Privacy, low-density urban environment, 176,
  183
Private spaces
  environmental sustainability, 179–181
  garden water use, 179–181
Property rights
  ecosystem management effects of, 270
  Imperial Valley, California case study, 270–
    271
  unarticulated assumptions, 270–271
Propositions, 54–55
Public gardens
  versus private gardens, water use, 179–181
  sustainability, 179–181
Public spaces
  Barcelona new urban lifestyle effects, 173–
    178
  collective versus individualistic influence,
    176–178
  direct observations, 165
  economic dimension, 169
  free accessibility property, 181–182
  integrated sustainability, 165–170
  in low-density cities, 177
    usage of, 183–184
    political dimension, 169–170, 184–186
  "methodological triangulation," 163–164
  multiple functions, 182
  openness property, 182
  political sustainability role, 161, 184–186
  properties, 181–184
  qualitative work, 164–165
  quantitative work, 163–164
  relational character of, 185–186
  "second modernity" impact on, 170–173,
    176–178
  service economy effects, 172
  social dimension, 161, 169, 182–183
  social sustainability, 181–186

Quality-of-life
  Austin, Texas new economy, 105–108
  Boston, Massachusetts new economy, 114–
    116
  British government's agenda, 209
  in integrated sustainability, 166
  in new economy spaces, 101–103
  South East England case study, 201
Quantification Settlement Agreement (QSA),
  280–281, 291

Radical urban political ecology, 56–61
Rambla del Raval, 138–140
Reading Borough Council study, 200–201, 209
"Real nappy" campaign, 85
Reclamation Act (1904), 290–291
Recycling, gendered mainstreaming, 85
Regional Spatial Strategy, England, 204–205,
  207

Regionalism. *See* New regionalism
Regulatory structure
    Imperial Valley water transfer, 272–295
    San Diego County water transfer, 289–293
    siloed form of, 289–290
"Remote control" model, 255
Rentiers, 143–144
Representative democracy, 230
Rescaling of the state, 143, 154n15
Risk society concept, 28–30, 202

Salinity
    Colorado River threat, 276
    control of, 276, 283
    Imperial Valley, California, 276
Salton Sea
    history, 275
    restoration plan, 281–283
    salinity problem, 276
San Diego County
    projected water consumption, 287
    water conservation, 288–290
    water transfer decision making, 287–290
San Diego County Water Authority (SDCWA)
    power of, 274
    and water conservation, 288–290
    water transfer decisions, 281, 283, 287–
        290
Scale, in conservation models, 257–258
"School run" congestion, 80
Science
    in Millennium Ecosystem Assessment, 51–
        52
    overconfidence in, 54–55
    and politics, need for integration, 54–55
"Second modernity," 170–173
Silicon valley, as pollution source, 97
Siloed regulatory structure, 289–290
Single family homes, water use, 289
"Smart growth," 4, 114
    ecological footprint effect, 206
    "Factor Four" theory link, 210n1
    neoliberal policy link, 209
    in new economy spaces, 114–115
    as paradigm shift, 206–207
    in South East England, 206
Social control
    and Barcelona Rambla del Raval, 138–139
    in politics of sustainability, 144–145, 155n19
Social issues
    collective versus private spaces, 178
    democracy as key to changes in, 35–38
    versus ecological issues, influence, 155n21
    and economic inequality, 221, 234n2
    gender inequality, 76–82
    and integrated sustainability, 165–170
    low-density urban model, 175
    and new regionalism, 221–223
    and public spaces, 161, 182–186
    in urban politics of sustainability, 145, 152,
        155n21
Social justice
    Barcelona, 137–139
    democracy as key to, 35–38
    entrepreneurial cities, 125
        food policy, 149–150

gender issues, 67–68, 76–82
    and integrated sustainability, 165–170
    politics of sustainability issue, 144–145,
        154n19
    in radical urban political ecology, 60–61
    sustainable development component, 98–99
Social power. *See* Power relations
Social relationships. *See* Social issues
Soil, in Imperial Valley, California, 276
South East England, 205–210
    case studies, 205–210
        methods, 194
    challenges to development in, 227–228
    community environment groups in, 200–
        201
    development rationalities, 224–229
    as economic "growth engine," 203
    environmental modernization, 201–202
    "institutional muddle" in, 206–208
    neoliberalism versus sustainable development,
        194
    new politics of development, 224–229
    "political modernization," 196–201
    Reading Borough Council study, 200–201
    regional government in, 203–210
    "smart growth" policy, 206–207
    social consumption considerations, 223–224
        infrastructure needs, 223
    spatial policy, 214–234
    Sustainable Communities Plan, 214–234
South East England Development Agency
        (SEEDA)
    central government role in, 205, 227
    purpose, 203
    "smart growth" initiative, 206–207, 210n1
South East England Regional Assembly
        (SEERA), 203–205
Southern California Metropolitan Water District
        (MWD)
    Colorado River water allocations, 277
    water delivery system, 274
    water transfer role, 279–283, 288
Spatial policy, 238–259
    Atlanta, Georgia, mobility crisis, 148–149
    in conservation planning, 238–259
    entrepreneurial cities, 145–150
    fragmentation model link, 244–256
    fragmented space in practice, 249–253
    governance, 226–232
    history, 225
    infrastructure concerns, 223
    and institutional power, 256–257
    Karvia, Finland case study, 250–251
    methodological and conceptual issues, 232–
        233
    new agendas, 229–232
    new regionalism impact, 221–223
    research directions, 233–234
    role of, 218
    social consumption factor, 223–224
    South East England, 214–237
        economic development emphasis, 224–
            229
    Sustainable Communities Plan, 214–234
    Vincent Drive England case study, 252–253
Special Areas of Conservation, 250

State control. *See* Central government
State Water Project, California, 274
"Story line" notion, 5–6
"Sub-politics," 29
Suburban sprawl
    environmental impact, 175–176
    Metropolitan Region of Barcelona, 172–173
    mobility politics, 148–149
    political effects, 148–149, 162, 187
    social effects, 162, 175, 187
    values and behavior in, 175
Sustainable Boston plan, 110–111
Sustainable communities
    agenda, 221–232
    consumption orientation, 224
    core features, 216–218
    development rationalities, 224–229
    discourse versus implementation in, 225–226
    early phases of, 215
    emergence of, 218–219
    in England, 214–234
    fundamental ecological limits in, 269–270
    governance, 226–228
    implementation deficits, 224–229
    methodological and conceptual issues, 232–233
    nation-states link to, 233
    new agendas, 229–232
    new politics of development, England, 224–229
    and new politics of space, 221–229
    new regionalism as product of, 221–223
    research directions, 233–234
    sustainability of, 229–232, 269–270
    twin discourses in, 219–221
Sustainable Communities Dialogue, 200–201
Sustainable Communities Plan, 69, 214
Sustainable Community Strategies, 198

Thames Gateway area, 225–226
"Third-way" program, 196, 202
Topological space
    in fragmentation model, 254–256
    "scalar abuse" in, 255
Toronto, urban metabolism, 58
Transportation
    Atlanta, Georgia case example, 148–149
    urban politics, 147–149
Tsunami, and environmental politics, 22
Turf grasses, irrigation, 180
22@ project, 133, 153n6

United Kingdom. *See also* England
    gender mainstreaming case example, 83–84
    sustainable communities agenda, 214–234
Urban development. *See also* Entrepreneurial city
    in new economy spaces, 95–122
    sustainability contradictions, 116–118
Urban environment
    Barcelona, compactness, 137
    conservation biology research, 246
    conservation planning, 251–253
        fragmentation model, 251–253
        Vincent Drive, England case study, 252–253

nonequilibrium ecology, 247
planning failures, 56
politicizing of, 55–61
as socioecological entity, 247

Urban governance
    collective provision politics, 146–147
    entrepreneurial cities, 125–131
    new economy spaces, 114–116
Urban lifestyles
    high-density mode, 174
    low-density mode, 175–176
Urban metabolism
    caveats, 58, 258
    definition, 58
    in radical urban political ecology, 57–59
Urban planning, failure of, 56
Urban sprawl
    environmental problems, 175–176
    Metropolitan Region of Barcelona, 172–173
    participative democracy effects, 187
    and politics, 162
    social effects, 162, 175

Vancouver, Canada, social justice, 149–150
Vincent Drive, England
    conservation planning case study, 252–253
    fragmentation model use, 252–253, 255

Waste management, gendered aspects, 79, 83–88
Water conservation, San Diego, 288
Water consumption
    low-density urban model, 176, 179–181
    in public spaces, 179–181
Water Resources Control Board, 279
Water transfer (Imperial Valley, California), 266–295
    agricultural effects, 273–278
    Colorado river allocations, 276
    commercialization of, 294
    decision-making process, 278–290
        environmentalists role, 291
    land development consequences, 283
    rural–urban relations, 273–275
    sustainability question, 286–287
    and water speculation, 283–284
Watershed planning
    and ecological models, 247
    in entrepreneurial cities, 144, 154n17
Welfare reform, sustainable communities, 220–221
Welsh Assembly, 193
Wildlands Project, 242
Wildlife corridors
    boundaries of 241
    as cultural construct, 244–245
    European Union plan, 250
    landscape ecology function, 241–242
    long-running debate, 244–246
    Vincent Drive, England case study, 252–253
Women's Environmental Network, 75
Women. *See* Gender inequality

Xeriscaping, 179–181

About the Editors

Rob Krueger is Assistant Professor of Geography at Worcester Polytechnic Institute in Worcester, Massachusetts. His research focuses on economy–environment relations in the contexts of urban/regional economic development.

David Gibbs is Professor of Human Geography and Director of the Graduate School at the University of Hull in the United Kingdom. His work focuses on local and regional economic development, with a particular interest in the use of environmental policy to support and inform economic development policies.

Contributors

Anna Batchelor, Centre for Human Geography, Brunel University, West London, United Kingdom

Susan Buckingham, Centre for Human Geography, Brunel University, West London, United Kingdom

James P. Evans, School of Environment and Development (Geography), Manchester University, Manchester, United Kingdom

David Gibbs, Department of Geography, University of Hull, Hull, United Kingdom

Andrew E. G. Jonas, Department of Geography, University of Hull, Hull, United Kingdom

Basil Katz, Department of History, University of Chicago, Chicago, Illinois

Roger Keil, The City Institute at York University, Toronto, Ontario, Canada

Rob Krueger, Interdisciplinary and Global Studies Division, Worcester Polytechnic Institute, Worcester, Massachusetts

Marc Parés, Departament de Geografia, Universitat Autònoma de Barcelona, Barcelona, Spain

Alan Patterson, Urban and Regional Studies, Sheffield Hallam University, Sheffield, United Kingdom

Stephanie Pincetl, Institute of the Environment, University of California at Los Angeles, Los Angeles, California

Mike Raco, Department of Geography, King's College London, University of London, London, United Kingdom

David Saurí, Departament de Geografia, Universitat Autònoma de Barcelona, Barcelona, Spain

Erik Swyngedouw, School of Environment and Development, University of Manchester, Manchester, United Kingdom

Aidan While, Department of Town and Regional Planning, University of Sheffield, Sheffield, United Kingdom